Power Generation in China: Research, Policy and Management

Chief Editors: Shi Dan, Yang Hongliang

Paths International Ltd

社会科学文献出版社
SOCIAL SCIENCES ACADEMIC PRESS (CHINA)

ADB Technical Assistance Project

(TA 7202-PRC)

Participants list (by alphabetical order):

Dong Jun, Li Pengfei, Shi Dan, Sun Yaowei,
Tan Zhongfu, Xia Xiaohua, Yang Jintian, Zhou Yuhui

Table of Contents

1. The Significance and Content of Energy-Efficient Power Generation Scheduling

The power industry is one of six high energy-consuming and high pollution industries. According to recent data in 2010, coal used for power generation accounted for more than half of the total coal use in the PRC responsible for 54% of sulfur dioxide emissions. While the water consumption of coal-fired power generation accounts for 40% of total industrial water utilization. In 2009, the electricity generation in the PRC consumes[1] 50 gce more coal than the international advanced level, which means around 180 million tons of standard coal equivalent was over consumed in that year, calculated with power output of 3681.2 billion kWh in the PRC[2]. In 2009, the line loss rate of PRC's power grid reached to 6.49%, which is 1% higher than the world's advanced level. It means more than 30 billion KWh of electricity was lost, equals 10.2 million tons standard coal equivalent.

In order to improve energy efficient, promote energy conservation, reduce environmental pollution, speed up adjustment of energy and power structure, ensure safety and efficient operation of power market and achieve sustainable development of power industry, the Office of State Council issued the "Regulation on Energy Conservation Power Generation Dispatching (draft)" (Office of State Council Document No. 53[2007]) in August 2007. It is jointly prepared by the NDRC and other three departments. The Approach requires reform of current power generation scheduling regime and carrying out the energy-efficient power generation scheduling. Pilot implementation of the new power generation scheduling was approved. And five provinces, namely Guangdong, Guizhou, Henan, Jiangsu and Sichuan, started testing the new scheduling. It has been proved that the implementation of energy-efficient power generation scheduling has important significance in reduction of energy consumption and pollutants emissions, as well as promoting of national economic development.

1.1 The significance of energy-efficient power generation scheduling

1.1.1 Energy-efficient power generation scheduling: to reduce coal consumption for power generation

In 2010, the PRC's total energy consumption accounted for 20% of the world's total

---

1   340gce/kWh.
2   Acculated according to the data issued by China Power.

amount, but GDP is 10% less than the world level; the PRC's per capita energy consumption is roughly equivalent to the world's average level, but per capita GDP is only 50% of the world's average level; the PRC's GDP is about the same as Japan, but the energy consumption is 4.7 times than Japan; total energy consumption in the PRC has surpassed the United States, while the economy is only 37% of the U.S[3]. In 2010, the PRC emitted 12.38 million tons of chemical oxygen, 21.85 million tons of sulfur dioxide emissions, ranking the first of the world, others such as nitrogen oxides, waste water and solid waste emissions are increasing rapidly.

Electricity, as a clean, high-quality and convenient secondary energy resource, is being widely used, accounting big proportion of total energy end-use. Between 1980 and 2009, the proportion of electricity in energy end-use in the PRC has been increasing, reached to 19.6% in 2009 from 6.8% in 1980. Power industry is a big energy consumer of energy while they produce or transfer energy. Compared with international advanced level, the coal consumption for thermal power generation in the PRC accounts is about 7.4% more than international advanced level; and coal consumption of electricity supply is 9.7% more than world's advanced level; also the line loss rate of power grid in the PRC still have 1.5% to go to meet the international advanced level. Improvement the proportion of electricity in energy end-use could improve the energy utilization rate, reduce energy consumption intensity, and further promote energy conservation. Energy-efficient power generation scheduling would improve the energy utilization rate in power industry, save energy and reduce pollutants emission.

Regarding the principle of power generation scheduling and the scheduling order, the new scheduling approach focuses on coal-fired power generation unites. Coal-dominated energy structure in the PRC results in the coal-based power structure. The proportion of coal-fired power generation accounts for more than 70% of total generated power. Besides, power source structure is irrational. Main problems include high energy consumption, low efficiency and utilization of small thermal power generation units which produce heavy pollution. According to the statistics, the PRC's coal consumption in power generation has accounted for more than 50% of total coal consumption, resulting in serious problem, including acid rain, ozone, fine particle pollution and greenhouse gas emissions. During the "11th Five-Year" period, to mitigate sulfur dioxide emission and acid rain pollution, the PRC has increased efforts on reorganization of power structure. The proportion of large

---

3 Liu Tienan: Rapid growth of China's total energy consumption highlighted the urgency of energy consumption control. July 9, 2011, Xinhua.com

fossil power generation units(300,000 kW and above) increased from 42.7% in 2000 to 69.4% in 2009; and 21 ultra-supercritical units with capacity of million or 10 million kW has been put into operation. However, in general, the overall structural problem of the PRC's power industry is still prominent, with high proportion of thermal power generation units (with capacity of 300,000 kW or below).

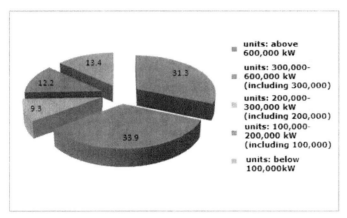

Figure 1-1   The proportion of installed capacity of thermal power generation units in 2008

As we may see from Figure1-1, by the end of 2008, the installation capacity (with capacity above 6000 kW) of thermal power reached to 586 million kW, of which, 31.3% is from units with capacity above 6000kW; 33.9% is from units with capacity between 300,000~600,000 kW (including 300,000); 34.8% is from units with capacity below 300,000kW. Different size of coal-fired generation units were significantly different in standard coal equivalent consumption, due to steam generating capacity, operating parameters and other operation efficiency is different. (See Table 1-1)

Table 1-1   Coal consumption of different generation units

Units	Coal consumption		
	g coal equivalent/(kWh)	g raw coal/(kWh)	10,000t/(100 million kWh)
Above 1 million kW	290	378	3.78
0.6~1 million kW (including 0.6 million)	300~320	420~448	4.20~4.48
0.3~0.6 million kW (including 0.3)	325~340	455~475	4.55~4.75
0.2~0.3 million kW (including 0.2)	360~365	504~511	5.04~5.11
0.1~0.2 million kW (including 0.1)	385~410	540~574	5.4~5.7
Below 0.1 million kW	490	686	6.86

As can be seen from Table 1-1, the coal consumption of large unit is lower, while the coal consumption of small unit is higher. As mentioned earlier, the implementation of energy-efficient power generation scheduling promoted the optimization of power source structure, as well as development of renewable energy generators and large-scale environmental friendly units.

1.1.2 Energy-efficient power generation scheduling: further optimization of power structure

Energy-efficient scheduling requires power grid companies to give the priority to renewable energy. The enforcement of the new rule will benefit renewable energy companies, and increase the proportion of renewable energy. According to statistics data, by the end of 2009, the proportion of installation capacity of non-fossil resources is increasing in the PRC. In total, the installation capacity (6000 kW and above) from the non-fossil energy reached to 222 million kW, of which, 196 million kW from hydropower, accounting for 22.46% of the total installed capacity. The proportion has been increased by 0.68% over the previous year. The PRC has become the world's largest country of hydropower installation capacity; 9.08 million kW from nuclear, ranking the ninth of the world; about 21.92 million kW of installation capacity is still under construction. The national wind power installation capacity reached to 17.6 million kW, 109.82% up over the previous year; in 2009, the national wind power generating capacity increased by 111.1%, higher than the growth rate of rest installed capacity. The integrated wind power installation capacity and net generation capacity has been double increased for four consecutive years. Figure 1-2 shows the trends of the power structure over years in the PRC.

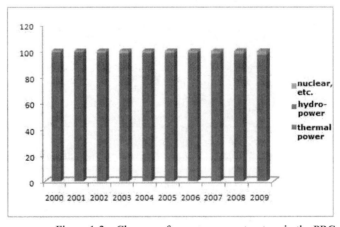

Figure 1-2   Changes of power source structure in the PRC

Meanwhile, with the implementation of energy-efficient scheduling, the utilization time of generating units with high energy consumption and high pollution will be reduced, while generators with high-performance be allocated for more generation assignments. Therefore, it will speed up the closing of power plants with lower efficiency and optimize the power structure. According to the statistics, during the 11th Five-Year Plan, the total capacity of decommissioned small fossil power plants reached to 76.82 million kW.

The implementation of energy-efficient power generation scheduling is conducive to emission reduction. It also helps to speed up the power structure adjustment, promote the development of renewable energy generators and environmental friendly units. So that inefficient small generation units will gradually be decommissioned. Which could promote the optimization and upgrading of overall power industrial structure.

1.1.3 Energy-efficient power generation scheduling: the way of reduction in air pollutants emissions

The implementation of energy-efficient power generation scheduling could not only promote the proportion of electricity generation from clean energy and renewable energy, but also help power source restructuring and reduction of fossil fuel consumption. Furthermore, after implementation of energy-efficient power generation scheduling, the utilization rate of high-efficient power generation units was significantly enhanced, while small-scale coal-fired units and fuel power generation units has difficulties to integrate to the grid. By limiting the small fossil power generation, abandon of small fossil power generators is encouraged to reduce the average coal consumption level of the power industry, and emissions of sulfur dioxide, soot, nitrogen oxides, Hg, carbon dioxide and other pollutants.

According to the estimated data from the National Development and Reform Commission in 2008, in the PRC, fossil power generation capacity of small generator with capacity below 100,000 kW accounts for nearly 30% of total power generation capacity. Small generators have problems such as, high energy consumption and heavy pollution. After the implementation of energy-efficient power generation scheduling, 90 million tons of standard coal equivalent would be saved annually. Which could significantly reduce emissions of sulfur dioxide, nitrogen oxides, soot, carbon dioxide, mercury and other pollutants, resulting in great environmental benefits (see Table 1-2).

Table 1-2 Environmental impact assessment for energy-efficient scheduling

Saved coal	Pollutant emission (10,000tons/year)				
10,000 tons standard coal equivalent/year	SO_2	NO_x	Flue gas	Hg	CO_2
9000	201	70.6	69.2	$2.37*10^{-4}$	32759.3

As can be seen from Table 1-2, after the implementation of energy-efficient power generation scheduling, 90 million tons of coal equivalent would be saved annually in the PRC, annually reducing 2.01 million tons of sulfur dioxide emissions, 706,000 tons of nitrogen oxide emissions, 692,000 tons of dust emissions, 2.37 tons of mercury emissions and 328 million tons of carbon dioxide emissions. Therefore, the energy-efficient power generation scheduling is conducive to better achieve the goal of energy conservation in power industry, to promote sustainable development in this field.

It can be concluded that energy-efficient power generation scheduling is a realistic choice to promote energy conservation. It is also an important measure to implement the scientific development concept of power industry. It addresses an institutional change for the original way of power generation dispatching. It is helpful to effectively reduce fossil energy consumption and pollution emissions, improve overall efficiency of generation units, and actively explore a sustainable development path for the PRC's power industry.

1.2 Primary approach of energy-efficient power generation scheduling

Power dispatch is the command center of power system. But the scheduling mode is determined by the scheduling system. Since the founding of new China, the power scheduling system in the PRC has been experienced a series of reforms along with the reform of micro-economic management system.

1.2.1 Related regulations on energy-efficient power generation scheduling

In August 2007, the State Council issued "Energy-Efficient Power Generation Scheduling Approach (trial)", which significantly impacted on reduction of energy consumption per unit GDP, transformation of national economic growth, promoting the reform of power industry and other aspects. Five provinces — Guangdong, Guizhou, Henan, Sichuan and Jiangsu, started testing the new rule. By implementation of new scheduling rule in pilot provinces, the new rule will be eventually applied nationwide aiming to all grid-connected power generation units.

In March 2008, the Electricity Regulatory Commission issued "Interim Measures for Regulation of the Trading of Generating Rights". The generating right trading is market oriented trading among generation units, power plants. The trading capacity includes contracted power capacity and power generating capacity indicators which issued by provincial governments. The principle of the trading is to promote replacement of low-efficiency high-pollution generation units with clean energy. Small generation units which are involved into the decommission program decommissioned on time or in advanced could conduct power generating rights trading according to related regulations.

April 3, 2008, the State Electricity Regulatory Commission, National Development and Reform Commission and The Ministry of Environmental Protection jointly issued "Information Dissemination Approach of Energy-Efficient Power Generation Scheduling (trial)". Three main problems are resolved, including, firstly, clarified the content of energy-efficient power generation scheduling information, and information which should be announced by the electricity regulatory agency, provincial related department, power dispatching agency and power plant; secondly, clarified the information dissemination method, time and objectives; thirdly: during the information dissemination period, the electricity regulatory agency and provincial government should play be functional with their own responsibility.

In December 2009, the State Electricity Regulatory Commission together with two other department issued "Notice Regarding on Economic Compensation in Energy-Efficient Power Generation Scheduling Pilots". The Notice required provincial government to conduct research and establish economic compensation approach for energy-efficient power generation scheduling in accordance with the actual situation. The Notice indicated that economic compensation should be prepared for generation units at cold reserve status. The Notice can be applied in all pilot provinces for generation units, and grid companies, of which generation units include all public units integrated to the main grid, units owned by power plants, and local units connected to the main grid.

1.2.2 Sorting method of energy-efficient power generation scheduling

Energy-efficient power generation scheduling is different from the traditional scheduling mode. It uses new scheduling rule instead of administrative allocation for power generation. Except independent power grid, power generation allocation of all grid-connected units would be decided by provincial and above power scheduling agencies. The ranking list of units is prepared according to unit type, energy consumption level and pollutants emission level. The ranking list of all generation units would be announced in advance to reasonable arranging backup capacity and strictly carry out safety check. Optimization of the combination mode of generation units should be conducted to achieve minimum energy consumption. During the real-time scheduling, scheduling should be carried out according to the power generation capacity and power generation hours in line with the announced ranking list.

The sorting table of generation units (hereinafter referred to as sorting table) is the main basis for energy-efficient power generation scheduling. The ranking list of provinces (autonomous regions and municipalities) is issued by the National Development and Reform Commission and instructed by the provincial people's government, and make timely adjustment according to unit capacity and the actual running time. The principles of ranking are described as below: 1) unadjustable wind, solar, oceanic, hydro and other renewable energy generators; 2) adjustable hydro, biomass, geothermal, and other renewable energy generators, as well as solid waste-fired units which meet environmental requirements; 3) nuclear power generation units; 4) coal-fired cogeneration units, generators which use waste heat, residual gas, residual pressure, coal gangue , washed coal, coal bed methane as the power resource; 5) natural gas, coal gasification based; 6) other coal-fired generation units, including cogeneration without heat load; 7) oil-based and oil product-based generation units. Energy conservation and emission reduction have been considered during preparation period of generation units ranking table. Units are ranked according to their energy efficiency. Units with the same energy efficiency are ranked according to their emission levels. Unit's energy consumption level uses the parameters provided by equipment manufacturers. However, the real-time measured values should be gradually used to replace the out-factory parameters. The increased coal consumption real-time value due to environmental protection and water conservation should be adjusted appropriately. Pollutants emission level should be in line with the latest issued level by provincial environmental protection departments.

The sorting table of each power source is prepared as below:

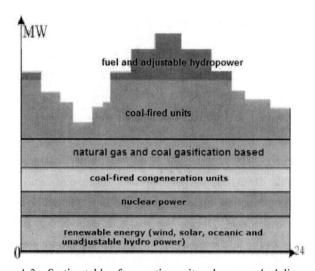

Figure 1-3   Sorting table of generation unit under new scheduling rule

As for the preparation of generating unit combinations program, the provincial Development and Reform Commission is responsible for the organization of annual, quarterly and monthly forecasting and management of electricity load requirement. They should also regularly release forecasting information to the relevant departments including the power grid companies and power generation companies. The annual, quarterly, monthly combination programme of generation units in provinces (autonomous regions and municipalities) was prepared according to the load forecast and actual operation situation of generation units. The ranking list of power generation combination program of different level of scheduling agencies is prepared according to daily electricity load forecasting and actual power generation capacity, power grid operation mode, as well as safety constraints and units on/off loss and other factors, to determine generator combination plan for the next day. The provincial power dispatching department determines the generator on/off mode in accordance with the ranking list of provinces (autonomous regions and municipalities) and the adjustable power generation capacity applied by units, to meet the power system security constraints, and report to the regional power dispatching department. Regional power dispatching department should further optimize and adjust the on/off mode of regional generators based on the provincial generator combination programme, in accordance with regional ranking list, adjustable power generation capacity, power delivery capacity of inter-provincial transmission line and network losses. Scheduling departments of State Grid Power Company and Southern Power Grid Company coordinates the generator's on/off mode and establish the generator combination programme for the next day based on power delivery capacity of inter-regional/provincial transmission line, network losses and sorting results, in line with the principle of Article 10 in the "Energy-efficient power generation scheduling Approach". This mode will be issued to the regional (provincial) executive power dispatching agency, and send a copy to provincial, regional and municipal Development and Reform Commission (Economic Commission), as well as the regional electricity supervising agency.

Different level of power generation scheduling agencies should reasonable allocate power generation load and prepare daily power generation curve according to following rules: 1) Renewable energy based units, except hydropower, power generation load is arranged according to the power curve submitted by power plants; 2) for unadjustable hydropower generation units, the principle of "power generation quotas are determined by the water amount" is adopted; 3) for those hydropower plants undertaking integrated utilization assignment, power generation load should be arranged under the precondition of meeting integrated utilization requirements, to increase hydropower utilization rate; for those cascade hydropower plants, reservoir optimization scheduling and combination dispatching of reservoirs should be carried out to rational use of reservoir; 4) for units of integrated resources utilization, the principle of "resources determine the power generation quotas" should be conducted; 5) unclear units, except few particular cases, the power generation quotas should be arranged according to the power curve; 6) for cogeneration of heat and power units, the principle of "heat determines the power generation quotas" should be adopted. Power load exceed the heat demands should be arranged as condensation units; 7) power generation load for thermal power generation units should be arranged in line with slightly increase rate under same coal consumption. The energy-efficient power generation scheduling should always put the safety as the first principle. Dispatching agencies should conduct safety review according to the requirements of "Guideline of security and stability of power system". Related decommissioning mode and adjustment of power generation quotas can be carried out to ensure stable operation of power system.

1.2.3 Basic procedure of energy-efficient power generation scheduling

Provincial Development and Reform Commission together with the provincial environmental protection department jointly review and approve the proposed ranking list submitted by power generation companies. The generation ranking list is prepared with consideration of generating unit type, energy consumption level, environmental protection indicators, configuration of water conservation facilities, and other factors. Declaration of new units should be in accordance with design parameters. Declaration of commerical generation units should be in line with actual real-time parameters. Parameters re-testing should be conducted after large-scale reparation and transformation. The ranking list should be issued only after public notice, and amended quarterly.

The provincial Development and Reform Commission prepares annual, quarterly, monthly power generating combination program in accordance with ranking list of generators and load forecasting results.

The provincial dispatching agency should determine generator combination of next day in accordance with related information reported by power companies, combined with the next day's power load forecasting, capacity plan of inter-provincial grid line, equipment maintenance, and security constraints, etc., as well as units ranking list and monthly generator combination program.

Regional dispatching agency should consider the provincial generator combination program and compare units' coal consumption, to achieve similarity of coal consumption, or capacity limits of inter-provincial power transmission. The basic procedure of energy-efficient power generation scheduling is shown in Figure1-4.

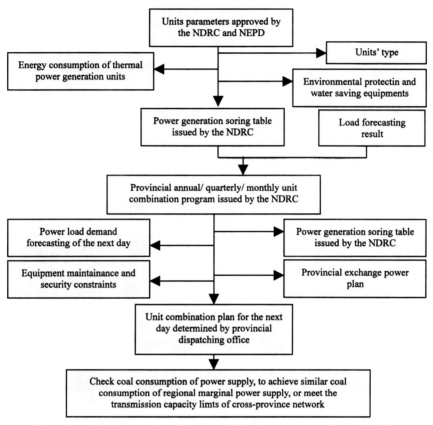

Figure 1-4   Basic procedure of energy-efficient power generation scheduling

1.3   Characteristics and impacts of energy-efficient power generation scheduling

1.3.1   Evolution of the power scheduling system in the PRC

(1) Planning dispatch

During the planning economy period, due to lack of power supply, many generation units are working together to meet the load demands. The government establishes the overall economic development indicators, power load forecasting amount and annual power generation plan. The power industrial department prepares the overall power generation arrangement. All power grids are unified. Back then, the power plant is only one workshop of power grid. The grid dispatching during that time is highly concentrated.

The Key System Factory (KSF) of grid dispatching in planning economic period is planned indicators. Related information of the planned indicator determines the power generation. The power plants and power grid companies closely cooperated. This scheduling rule is in line with the planning economy system.

(2) Economic dispatch

Due to the power shortage in 1970s-1980s, governments encouraged to investment in power plants construction. Many generation units with different fuel transition characteristics and different ownership are introduced to power generation. Related economic problems came along with balancing the interests of power plants and power grid companies to save cost and reduce energy consumption. Those problems mainly include how to arrange load allocation for companies have many units, and how to minimize fuel consumption. Therefore, the economic scheduling is established.

The economic scheduling aims to optimize capacity allocation of units based on their energy consumption level. The dispatching principle adopted back then is slightly increase rate under same energy consumption level. The key point of economic dispatching is to achieve minimum coal consumption at regional power grid. The power generation load is allocated according to relations curve of input and output power. The economic scheduling was only implemented in some regions. It is hard to be applied at a large area since we have the problem of continuous shortage of power supply even during the valley period. Besides, policies and regulations of economic scheduling are imperfect. For example, power generation quotas have to be allocated to some power plants to balance interests of power plants.

(3) Grid integration auction

In 2002, the State Council launched the reform program of power system, which is "separating power plants and grid companies, auction for grid integration". The new program allows to sale the power generation assets of State Power Corporation and establish five power generation companies (Huaneng, Datang, Guodian, Huadian and China Power Investment Corporation), State Grid, China Southern Power Grid through restructuring and mergers. Few years later, pilots of auction for grid integration carried out in power markets of northeast and east China. The electricity price includes two-settlement price model and one-part price model. Currently, the pilot projects have been completed. The resource allocation in power market is not realized through market mechanism. Power generation quota is allocated through planning mechanism according to installed capacity of generation units in power plants. The equipments utilization hour is even among power plants. The on-grid electricity price is based on the cost.

(4) Energy-efficient power generation scheduling

Since the Chinese government first proposed emission reduction goals in the "11th Five-Year Plan", under the guidance of the government, new rules for the implementation of grid scheduling – Energy-efficient power generation scheduling.

The Key System Factory of energy-efficient power generation scheduling is that ranking list of generating units is determined by units' efficient performance. All units can be ordered as: renewable energy based units, including hydro, wind, biomass and solid waste-based power generation; nuclear; natural gas, including coal bed methane, gasification based, heat cogeneration; coal-fired generation; oil-based generation. The new scheduling rule gives priority to renewable and clean energy under the precondition of rational dispatching available power supply equipments in related regions. It aims to reduce environmental pollution in an efficient and environmental-friendly power generation manner. The implementation procedure of the new scheduling rule is: forecasting next-day's load demands, inter-provincial power generation quotas exchange plan, equipments maintenance and safety constraints. The next-day's generating units combination program is determined according to the units ranking list provided by provincial Development and Reform Commission, to achieve energy conservation and environmental protection goals.

1.3.2 Comparison of different generation dispatching modes

(1) Energy-efficient power generation scheduling and economic dispatching

The goal of energy-efficient power generation scheduling is energy conservation and environmental protection, to achieve the optimization and improvement of power system. It is an important measure to achieve energy conservation at the primary electricity market environment.

Compared with economic dispatching, energy-efficient power generation scheduling has been improved in the following areas: 1) the "Approach" clearly defined "priority dispatch of renewable generation resources," which gives priority to renewable energy generation to access the grid from the perspective of institutional arrangements. Also encouragement policies were prepared for renewable generation resources; 2) generator sorting considered i energy conservation, giving priority to lower energy consumption units. If they consume the same amount of energy, then give priority to less pollutant emission ones; 3) power dispatch agencies should actively carry out hydropower joint optimal dispatching and optimal dispatching of fossil power and hydropower, to improve utilization rate of water resource, and maximize peaking value of hydropower; 4) combined with the basic situation of the PRC's electricity market, take the power generation dispatching as an opportunity to gradually achieve the transition of primary means from administrative arrangements to market oriented mechanisms.

Table 1-3 Comparison of dispatching modes

Scheduling mode	Objectives	Constraints	Characteristics
Traditional economic dispatching	Least fuel consumption or cost	Max. and min. output power of balanced units	Slight increasing rate of coal consumption
Auction model	Least coast of power purchasing	Power plant is separated with power grid, network security, grid power balance	Optimal grid power purchasing
Energy-efficient power generation scheduling	Energy conservation, environmental friendly, economic benefits	Power balance, complicated constraints of grid and generation units and environmental capacity constraints	Integrated goals

Energy-efficient power generation scheduling is a combination of administrative measures and market mechanisms, reflecting both the technical optimization and economic compensation. The new scheduling approach pays more attention on energy conservation and environmental protection, with consideration of the market mechanism and administrative measure. In order to successfully implement energy-efficient power generation scheduling, in addition to a reasonable sort of generators, optimizing the scheduling mode and economic compensation measures, the market mechanism should also be explored to achieve a well-organized electricity market. Table 1-3 shows the comparison of different scheduling modes.

(2) Energy-efficient power generation scheduling and grid integration

The energy-efficient power generation scheduling mainly considered energy consumption and the pollutant emissions indicators, as well as determining the online ordering of generators, preparation of generator combination program and unit load allocation. Bidding strategies is based on the reported level of tariff to determine the order of power generation. In the power market, dispatching priority has been given to renewable energy unit, as for coal-fire power generation units; the price mechanism is the dominant of consideration.

Both energy-efficient power generation scheduling and grid integration auction aimed to reduce energy consumption and rational use of energy. However, energy-efficient power generation scheduling pays more attention on decreasing energy consumption and reducing pollutant emissions, while the grid integration auction pays more attention on the optimal allocation of resources by using market mechanism. From the price point of view, there are significantly different. The price system used by energy-efficient power generation scheduling is approved by the government, while price system used in the power market includes contract price and spot price. Considering the power industry's sustainable development, they are the same thing. Energy-efficient power generation scheduling integrated with auction system can promote the establishment of market-oriented operation mechanism for power industry, enhance the generation rate of efficient units, guide the inefficient small fossil power generation units to decommission from the market, to further optimize resource allocation.

1.3.3   Impacts of energy-efficient power generation scheduling on the PRC's power system

The impacts on power grid dispatching due to energy-efficient power generation scheduling can be summarized as below:

(1) Changed the dispatching rule from even allocation of power generation quotas to units ranking by energy consumption level

The energy-efficient power generation scheduling gives the priority to units with high efficiency and low energy consumption. It changed the traditional even allocation mode for power generation quotas. It is helpful for energy conservation. The direct impact on power plants due to implementation of the new scheduling rule is the technological level of generating units directly determines their profits. It might have some slight impacts on personnel adjustment in power plants. For the new rule, the power generation technology is getting extremely important. The implementation of the new scheduling rule would promote the technological reform of power plants to reduce coal consumption and improve competition capacity.

(2) Optimized units combination program

The new scheduling rule considered the energy conservation and environmental protection factors by arranging resource allocation according to energy consumption level. The power dispatching center prepares the ranking list of generating units and units combination program in accordance with energy-efficient power generation scheduling.

The new scheduling rule gives priority to renewable energy based units, followed by clean energy, including natural gas, clean coal-fired power generation. The medium and small units would only be used during the peak period if the supply cannot meet the demands. Therefore, after the implementation of energy-efficient power generation scheduling, changes are not only happened to the ranking position of units, but also the units combination program. Take the hydropower generating units as example, before the implementation of new scheduling rule, they are backups. But now, they have priority for power generation. Meanwhile, the utilization hour of units is depended on coal consumption level. Units with high coal consumption level would have less utilization hours. Few might be utilized less than 100 hours during the peak period. The experience in Sichuan pilot shows that those units with high coal consumption level are facing a bad situation, either losing money or being decommissioned soon.

(3) Impacts on operation cost of power grid companies

For those power grids which are used to large capacity coal-fired units, the new scheduling rule might impact on their operation management and safety maintenance due to introduction of renewable energy. Secondly, the renewable energy based power price is favorable determined by the government, which could increase the power purchasing cost to power grid companies. For example, the grid-connected power price of the coal-fired power plant is CNY0.38/kWh; the gas-based power price is CNY0.48/kWh; grid-connected power price for wind power is 0.58~0.68/kWh; the solar power price is 1.09/kWh. It means the power price is higher for those units have the purchasing priorities. Besides, the power grid companies should also buy units with ancillary service. Thirdly, renewable energy like wind and solar power would impact the stability of power grid and power generation quality. The dispatching planning of power grid has to be adjusted or reformed. Calculated with data from Sichuan pilot, the average power purchasing price is increased by 3% under energy-efficient power generation scheduling. However, the integration of new energy is a necessity soon or later. If it happened soon, it would bring more opportunities for national energy technology improvement. If it happened late, we could only follow the development of renewable energy in developed countries.

(4) Promote fundamental reform of power resource planning

The energy-efficient power generation scheduling, for one side, depends on the participation qualifications of power plants for grid-connected power generation; for another side, it establishes a new rule of market competition for power plants. Therefore, it is a new market rule targeting to energy conservation. At present, power plants cannot choose their favorable price system, market and power purchasers under current market mechanism of power industry. The new scheduling rule is the lifeline to determine how much power generation quotas and how much profit can be made by power plants. Therefore, under this rule, the power plants should fully consider the type of generating units, their capacities, power plants location, grid-connection mode, and unit combination program and pollutants emissions levels. Apparently, the energy-efficient power generation scheduling could adjust the investment direction and scale of future power industry. After two years implementation of the new rule, in March 2010, China Huadian Group announced to construct Laizhou power plant with millions kW capacity. It would be a high-efficient, low-consumption and zero sewage discharge power plant.

(5) Adjusted power grid planning and power industrial layout

The power grid planning is the pre-arrangement of constructions in power supplying regions. Power grid planning problems might be greatly changed, including balance between power transmission and power distribution network, grid-connection mode, changes of various voltage lines, reconstruction of power transmission equipments, investment scale and time, micro-power grid planning, renewable power source layout, available installed capacity and so on.

The new scheduling rule would cause great impacts on power industrial layout. The renewable energy based power grid would always be considered at the first position under the new rule. The technical parameters of power generation capacity should be designed under the precondition of ensuring the supply-demand balance of coal-fired units, as well as the balance between active trading and ancillary service.

Overall, the planning dispatch, economic dispatch, energy-efficient power generation scheduling are different modes under different power management regime and market environment. The energy-efficient power generation scheduling promoted the demand-supply balance theory in power system and the experimental innovation. It forms the characteristics of operation modes in PRC's power system. It has been noticed by global power industries.

2. Pilots Analysis of Energy-Efficient Power Generation Scheduling

2.1 Effect analysis in Guizhou province

2.1.1 Basic situation of Guizhou Power Grid

(1) Primary introduction

Guizhou province is situated in the eastern part of southwest region in the PRC. The Guizhou Power Grid belongs to China Southern Power Grid, connected with Sichuan, Chongqing, Yunnan, Guangxi, Guangdong Power Grid, covering 176,000km², population of 37.93 million, accounting for 2.86% of total population in the PRC.

By the end of April 2010, the installed capacity of Guizhou Power Grid reached to 25.67million kW, including 16.54million kW installed capacity of thermal power and 9.13 million of kW hydropower. In 2009, in total 117.1 billion kWh on-grid electricity has been dispatched by Guizhou Power Grid, of which thermal power reached to 96.324 billion kWh, and hydropower reached to 20.776 billion kWh. About 63.7 billion kWh electricity has been sold within Guizhou Province, and 44.2 billion kWh has been transmitted out of Guizhou.

(2) Power balance prediction

1) Dispatched load and power consumption prediction

The increase rate of electricity utilization in Guizhou was between 9% and 12% during the later period of the "11th Five-Year Plan". It is estimated that the electricity demand in Guizhou in 2020 will reach to 137 billion kWh with consideration of its rich mineral and coal resources, as well as integrated power consumption and electric power elasticity factor. The 12th Five-Year Plan of China Southern Power Grid proposed three electricity demands plans with different levels for the 12th Five-Year Plan and medium-long term period in Guizhou. It forecasted the maximum load would reach to 14 million kW, 14.9 million kW and 15.7 million kW in low, medium and high demand respectively.

The maximum load in the winter in Guizhou normally happens around 8 o'clock in the evening, while the maximum load in the summer normally happens around 9 o'clock in the evening. In 2006, daily load rate in winter and summer reached to 0.86, and daily minimum load rate was 0.743 in the winter and 0.7 in the summer. Along with increasing demands of electricity in the tertiary industry and residential use, the daily load rate and minimum load rate will be decreased. According to the forecasting results, between 2010 and 2020, the daily load rate is 0.84~0.81, daily minimum load rate in the summer is 0.686~0.65, daily load rate in the winter is 0.657~0.62.

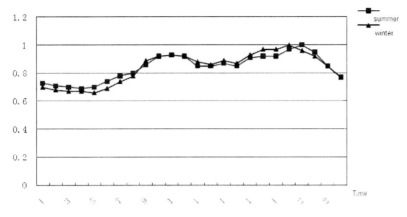

Figure 2-1   Typical daily load prediction curve in Guizhou

The maximum load in Guizhou normally happened in November and December, which is the annual peak period of power utilization. The sub-peak happened in March and April, about 87% and 90% of that is in November and December. May and June is the valley period except spring festival, due to cool weather in May and June and power generation from small hydropower station. The electricity demand increased after July and reached to the maximum in November and December. The change rule of annual maximum load in Guizhou would not be changed in the short term due to its geography, weather, transportation situation and living habit in Guizhou province.

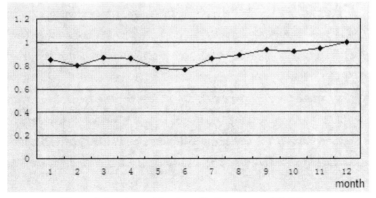

Figure 2-2   Annual load prediction curve of Guizhou

Data source: Scheduling operation summary of China Southern Power Grid in 200

2) On-grid power generation forecasting

In the 11th Five-Year Plan, Guizhou plans to invest for 16.2 million kW, including 7.3 million kW of hydropower, 8.9 million of coal-fired power. The capacity of small thermal power generating units to be decommissioned will reach to 230,000 kW.

The basic principle of power construction in the "12th Five-Year Plan" of Guizhou is to actively develop hydropower and optimize thermal power. Priority will be given to hydropower, especially to development of Wu River and Beipan River basin. According to the distribution of coal and its development situation in Guizhou, layout of coal-fired power should be optimized, as well as pithead power plants. We should further promote decommission of small thermal power plants to reasonable allocate local power resources.

In 2010, considering the capacity of decommissioned units would reach to 400,000 kW in Pan County, in order to meet the power demand at medium load level in Guizhou, about 1.8 million kW thermal power would be newly put into operation. However, to meet the high load level, about 2.4 million kW thermal power resources would be needed. The total installed capacity in Guizhou in 2010 reached to 31.19 million kW, including 11.18 million kW of wind power (36%) and 20.01 million kW of thermal power (64%).

3) Power transmission to outside of Guizhou

During the 11th Five-Year Plan, three thermal power projects are planned to be established by Guizhou and other neighboring provinces, including Xingyi Power Plant (power transmission to Guangxi province), Xishuierlang Power Plant (to Chongqing) and Qiandong Power Plant (to Hunan). The total installed capacity would reach to 3.6 million kW. Electricity generated by those power plants would be sold to those partner provinces/cities.

In line with national "west-to-east power transmission", Guizhou province will transmit 4 million kW of electricity to Guangdong province during the 11th Five-Year Plan. Meanwhile, according to the forecast of possible power demand from neighboring provinces, Guizhou will transmit 10 million kW of electricity to neighboring provinces/regions in the next few years, including 8 million kW to Guangdong province and 2 million kW to other neighboring provinces.

According to the demand prediction, in 2010, the power demand within Guizhou would reach to 12.91 billion kW, plus 10 million kW of electricity involved into the "west-to-east power transmission". Considering the reserve capacity should be 15%, the total installed capacity in Guizhou Grid should reach 26.34 million kW.

2.1.2 Measures adopted in Guizhou

After nearly one year preparation and simulation operation, the pilot of energy-efficient power generation scheduling was established in Guizhou on December 30, 2007.

(1) Simulation operation program of energy-efficient power generation scheduling

December 1-31, 2007, energy-efficient power generation scheduling of Guizhou Power Grid carried out the simulation operation. The core work of simulation operation includes the procedure of preparation and implementation of energy-efficient power generation planning curve and procedure of information dissemination of energy-efficient power generation scheduling.

(2) Implementation principles of energy-efficient power generation scheduling in Guizhou

The "Implementation Principles of Energy-Efficient Power Generation Scheduling in Guizhou" clearly defined specific operational approaches of power dispatching in Guizhou, including power generation ranking list, load forecast and generation units combination program, preparation of daily power generation planning, safety check, peak shaving and frequency regulation and backup, emission monitoring and other operational approaches. Under the new rule of energy efficient power generation scheduling, all grid-connected generation units in Guizhou are classified into the following priority categories: 1) unadjustable renewable resource, including wind power and adjustable hydropower; 2) adjustable renewable resource, including hydropower and biomass. During the flooding period, priority should give to adjustable hydropower generation units; 3) residual-heat, residual-gas, residual-pressure solar, coal gangue, middling and coal bed methane; 4) coal-fired generating units.

By September 1, all power plants should provide real-time measured value of on-grid generation units in terms of parameters of energy-efficient power generation scheduling, and design value of on-grid generation units for the next year under the new scheduling rule to provincial Economic and Trade Commission (or NRDC). Those values will be submitted to designated agency by provincial Economic and Trade Commission (or NRDC).

By October 31, provincial Economic and Trade Commission (or NRDC) should prepare sorting table of generation units and allocate to related scheduling agencies, as well as release information to the public, with consideration of units' type, energy consumption level, environmental issue and water conservation factors.

The power scheduling agency in Guizhou prepares the daily power generation combination program according to daily load forecast and actual power generation capacity of units, with consideration of grid security, unit on/off, equipment commissioning and maintenance, grid loss and other factors. They also are responsible for the quotas of power load allocated to ongoing generating units.

All on-grid generating units should be involved into peak shaving and frequency regulation and backup according to dispatch instructions. In order to ensure power quality and grid security, generation units providing ancillary services should be compensated in accord with "Evaluation management regulation on scheduling of power plants in Guizhou". The adjustable hydropower units and coal-fired units are normally responsible for peak shaving. The capacity of peak shaving for thermal power generating units should not be less than 50% of rated capacity.

In terms of monitoring on pollutant emission, flue gas on-line monitoring system is installed in coal-fired generating units in Guizhou. And it is under dynamic supervision of power scheduling agency in Guizhou.

(3) Workload arrangement of energy-efficient power generation scheduling in Guizhou

In December 2007, the National Development and Reform Commission approved the "Workload arrangement of energy-efficient power generation scheduling in pilot areas in Guizhou". The dispatching work is conducted by Guizhou Provincial Economic and Trade Commission. Detailed workload planning and progress table is established, to carry out the work in three stages.

(4) Information dissemination approaches of energy-efficient power generation scheduling in Guizhou

Guiyang Electricity Regulatory Commission Office established "Information Dissemination Approach of Energy-Efficient Power Generation Scheduling in Guizhou" clearly clarified that, Guizhou Provincial Economic and Trade Commission should be responsible of providing information dissemination content and time to Guizhou grid dispatching agency; power generation companies shall provide information dissemination content and time to Guizhou Provincial Economic and Trade Commission and Guizhou grid dispatching agency; Guizhou grid dispatching agency should be responsible of dispatching content and time release.

(5) Operation regulation for desulfurization online monitoring system of on-grid coal-fired power generation units in Guizhou

Desulfurization on-line monitoring system includes desulfurization real-time monitoring system of coal-fired generation units and desulfurization information dissemination and management system. The "Operational management regulation for desulfurization online monitoring system of integrated coal-fire power generation units in Guizhou" described the responsibilities of related agencies and departments during construction, operation, maintenance, management of desulfurization online monitoring system.

(6) Operation and management regulation for coal consumption online monitoring

system of energy-saving power dispatching in Guizhou

The operation and management of coal consumption online monitoring system includes daily operation management, security management, equipments evaluation and defect management, equipment maintenance and re-served management, fault handling and accident repair management, spare parts and technical data management, and indicators management of operational statistical analysis. As for construction, operation and management of coal consumption online system, it requires that coal-fired generation units with capacity of 200,000 kW must establish coal consumption online monitoring system; and units with capacity below 200,000 kW should consider duration of service, be voluntary established.

2.1.3 Achievements of energy-efficient power generation scheduling in Guizhou

Guizhou is the first pilot in the PRC to operate energy-efficient power generation scheduling. It successful implemented energy-efficient power generation scheduling of optimized hydro-thermal energy, earlier studied and established coal consumption online monitoring system to be applied in actual generation sequencing. At the first day of implementation of energy-efficient power generation scheduling, about 592 tons of coal equivalent was saved.

In 2009, the integrated line loss rate of Guizhou Power Grid Company reached to 5.7%, declined by 0.2% compared with 2008 (5.9%); coal consumption of dispatched thermal power plants for power generation reached to 322 g coal equivalent/kWh, decreased by 6 g coal equivalent/kWh compared with 2008; coal consumption of dispatched thermal power plants for power supply reached to 347 g coal equivalent/kWh, decreased by 5 g coal equivalent/kWh compared with 2008.

Comparison of power generation in 2009 and 2008 is shown in Table 2-1.

Table 2-1 Thermal power generation situation of Guizhou province in 2008~2009

Project and time	2009	2008	Compared with previous period (%)
Coal equivalent consumption for power generation(10,000t)	322	327	−1.53
Coal equivalent consumption for power supply(10,000t)	347	353	−1.70
Power consumption rate(%)	5.60	5.80	−3.45

Data source

During the implementation of energy-efficient scheduling between December 2007 and December 2010, Guizhou has established a leading power generation energy-saving system which includes organizational system, technology and management system. A total of 3.09 million tons coal has been saved, equivalent to reducing 9.5 million tons of CO_2

emissions, 188,900 tons of SO$_2$ emissions. The provincial average of desulfurization rate for thermal power plants reached to 95.4%, to reduced 3.6 million tons of SO$_2$ emissions. Energy-efficient power generation scheduling effectively promoted the energy conservation of power industry. In the first three quarters of 2010, the energy consumption decreasing rate of GDP in Guizhou reached to 4.39%, ranking the 2nd place in the PRC.

Guizhou province established and improved a series of policies and measures related to energy-efficient power generation scheduling, providing policy supporting to pilot work. The technical system of energy-efficient power generation scheduling has been completed, providing reference to implementation, establishment, operation and evaluation of each dispatching system.

(1) At policy level

Considered the actual situation of each pilot in Guizhou province, seventeen regulations and technical standards have been prepared covered management, operation and evaluation, mainly including "Simulation Operation Program of EEPGS", "Implementation Specification of EEPGS", "Pilot Work Program of EEPGS in Guizhou", "Sorting Table of Thermal Units for EEPGS in Guizhou", "Information Dissemination Approach of EEPGS in Guizhou", "Flue Gas Desulfurization Online Monitoring System Operation and Regulatory Approach of On-grid Coal-fired Units in Guizhou", "Operation Management Regulations on Flue Gas Desulfurization Online Monitoring System of On-grid Coal-fired Unit in Guizhou", "Coal Consumption Monitoring System Management Approach in Guizhou" and "Operation Management Regulations on EEPGS Coal Consumption Online Monitoring System in Guizhou".

(2) At technical level

In the end of 2008, Guizhou Power Dispatch and Communication Bureau completed the development of integrated system of energy-efficient power generation scheduling in Guizhou. This system consists of generation system of energy-efficient power generation scheduling for Guizhou Power Grid, information dissemination system of energy-efficient power generation scheduling, flue gas desulfurization online monitoring system for on-grid coal-fired units, coal consumption online monitoring system of energy-efficient power generation scheduling, and energy-efficient power generation scheduling system considering security constraint and network loss modification, providing technical supports.

1) Generation system of energy-efficient power generation scheduling for Guizhou Power Grid

Firstly, this system could read the power load forecasting results as the basis of power generation plan. Then it could automatically exclude units in maintenance and backup units.

The thermal generating units are sorted according to the "Sorting table of on-grid thermal power generating units in Guizhou". The sorting table is prepared according to the pollutant emission, energy consumption level. It optimized the thermal power units, giving priority to thermal units using desulfurization equipment and consuming less energy, to reduce coal consumption.

2) Flue gas desulfurization online monitoring system for on-grid coal-fired units in Guizhou

The flue gas desulfurization online monitoring system for on-grid coal-fired units in Guizhou was put into effective in July 2007. It is the first time in the PRC to realize auto-tracking, statistical analysis, report generation, information dissemination and query applications through systematically analysis and calculation by online operation time of coal-fired units, operation time of desulfurization equipments, operation time ratio, effective desulfurization power, daily SO_2 emission amount and other operation indicators.

3) Coal consumption online monitoring system of energy-efficient power generation scheduling in Guizhou

In April 2008, the main body of coal consumption online monitoring system of energy-efficient power generation scheduling in Guizhou completed. And it was applied and put effective in Dafang power plant in data access, debugging, data checking. This system collects the real-time operation data of generating units, calculates real-time efficiency and energy consumption of units, to provide real-time and reliable data for new scheduling rule. In total, 58 generating units from 17 power plants have been involved into energy-efficient power generation scheduling. In May 2010, it is the first time in the PRC that power generation scheduling in Guizhou is carried out according to the sorting table which prepared by real-time coal consumption value.

4) Energy-efficient power generation scheduling system considering security constraint and network loss modification

The main function of the energy-efficient power generation scheduling system considering security constraint and network loss modification includes trendy technology, security check of power generation planning, security constraint of network loss, optimization of power generation planning, network loss modification of energy-efficient power generation scheduling, to reduce network loss and coal consumption.

2.1.4 Existing problems and development trends of energy-efficient power generation scheduling in pilot areas in Guizhou

Main existing problems of energy-efficient power generation scheduling in Guizhou can be summarized as following three aspects:

(1) Economic compensation

Aiming to generating unit combination program and power generation plants which actual power generating hour is lower than limitation standard, related supporting policy, countermeasure and effective economic compensation mechanism should be prepared to incentive related power generation companies, to ensure reasonable reserve capacity.

(2) Coordination problems between power generation dispatching and large users as well as direct trade among power generating companies

In April 2009, Guizhou launched the direct trading policy between large users and power generating companies, which aims to minimize the cost of purchasing electricity. The implementation of direct trading has impacted the sorting table of thermal power plants, deviated from the principles and objectives of energy-efficient power generation scheduling. How to achieve well convergence between large users and power generating companies is a problem faced by energy-efficient power generation scheduling.

(3) Insufficient of primary energy

In 2010, coal supply for thermal power plants is insufficient, generation units with large capacity and high efficiency was decommissioned due to lack of coal, which cause serious impact on energy-efficient power generation scheduling work.

2.2 Effect analysis of energy-efficient power generation scheduling in Guangdong

2.2.1 Basic situation of Guangdong Power Grid

The power system of Guangdong province includes not only the "west-to-east power transmission", "three gorges to Guangdong" and other inter-provincial networks, but also cross-regional networks to transmit power to Hongkong and Macau. The structure of power network is complicated covering all types of power generation units. Meanwhile, the workload of peak shaving of Guangdong Power Grid is heavy, and the system operation is complicated. As a pilot place, this province has a certain representation.

Guangdong Power Grid is the biggest provincial power grid in the PRC. By April 30, 2010, the commercial used power reached to 51.44 million kW by Guangdong Power Grid (including 1.2 million kW dispatched from China Southern Power Grid). The installed capacity within local dispatch reached to 5.894 million kW. As for the central dispatch, in total, 34.242 million kW of coal-fired units, 7.456 million kW of gas-fired units, 1.92 million kW of oil-fired units, 3.948 million kW of nuclear units, 875,000 kW of hydropower units and 3 million kW of storage units; as for the districted dispatch, in total, 3 million kW of thermal power units, 2.224 million kW of hydropower units, 662,000 kW wind power and waste-based units. In 2009, the highest dispatched load reached to 63.608 million kW, highest load of west power reached to 20.709 million kW, and highest load of Hongkong power reached to 1.802 million kW in Guangdong province.

2.2.2 Measures adopted in Guangdong for EEPGS

On November 17, 2008, Guangdong Province launched the pilot program of energy-efficient power generation scheduling, followed by Guizhou province. Working mechanism and implementation plan was established to further improve technology systems and supporting policies. Work program of energy-efficient power generation scheduling in 2008 was proposed, as well as energy conservation management approach of Guangdong Power Grid Company, operation and management regulation of energy-efficient power generation scheduling in Guangdong power industry, management regulation of information dissemination website of energy-efficient power generation scheduling for Guangdong Power Grid Company.

Guangdong Province has developed a series of power generation dispatching policies, and developed technical supporting system based on "Sorting Table for EEPGS in Guangdong Province". At the policy level, Guangdong Province established "Pilot Implementation Program of EEPGS in Guangdong Province". At the technology level, based on the "Sorting Table for EEPGS in Guangdong Province", study on daily dispatching planning system was completed; technology supports and information management system platform for dispatching was established, as well as desulfurization online monitoring system of integrated coal-fire power generation units and cogeneration units' heat load online monitoring system, to achieve real-time monitoring for emissions and cogeneration units; information dissemination website for energy-efficient power generation scheduling was established, so that all information can be accessed by this platform.

Guangdong province prepared a series of policies for energy-efficient power generation scheduling, and developed technical supporting system based on the "Sorting Table of EEPGS in Guangdong province".

At the policy level, Guangdong province issued "Implementation Program of EEPGS Pilot Work in Guangdong Province". It detailed the workload of pilot places, clarified main work content, progress and responsibilities. Also, the sorting table of generation units was prepared, as well as unit combination program. The information dissemination approach of energy-efficient power generation scheduling was prepared to identify the content, time, scope and release channel of energy-efficient power generation scheduling information. The work implementation program was prepared to integration of promote local generation units to provincial dispatch agency. Except Shenzhen, by the end of 2009, in total 12 local power plants which installed capacity above 100,000 kW, have been integrated to the provincial power dispatch agency in Guangdong province.

At the technical level, based on the "Sorting Table of EEPGS in Guangdong province", study on daily dispatching planning system was completed, and information management platform consisting technical supporting system and information management system of energy-efficient power generation scheduling was established, as well as desulfurization online monitoring system of grid-connected coal-fired power generation units and cogeneration units' heat load online monitoring system, to achieve real-time monitoring for emissions and cogeneration units. Besides, information dissemination website for energy-efficient power generation scheduling was established, so that all information can be easily accessed by this platform.

The assistant technology supporting and information management system of energy-efficient power generation in Guangdong province covers establishment of scheduling plan, evaluation of energy conservation benefits, system load forecasting, bus load forecasting, net loss analysis and correction, security check and other functions. The preparation of power generation planning becomes more scientific, the management of energy-efficient power generation scheduling becomes more detailed, to achieve security, energy conservation and environmental protection in power grid scheduling.

The desulfurization real-time monitoring system of coal-fired units achieved functions, such as real-time operation status monitoring of desulfurization equipment, operation time and pollutant emissions (like SO_2), to meet technical requirements of energy-efficient power generation scheduling. By the end of 2009, in total, flue gas desulfurization information of 108 coal-fired generation units has been integrated to Energy Management System (EMS) of Guangdong provincial power scheduling center.

Through installation of heat load real-time monitoring system for heat cogeneration units, heat load information of five heat cogeneration power plants (Shuangshui, Maoming, Hengyun, Huarun Huangge, Nanhai power plants) has been integrated to EMS of Guangdong provincial power scheduling center.

Information dissemination system of energy-efficient power generation scheduling achieved information share among Southern Electricity Regulatory Agency, Provincial Development and Reform Commission, Provincial Economic and Trade Commission, Provincial Environmental Protection Department, Power Scheduling Agency and Power

Plants. It is the first information dissemination system of energy-efficient power generation scheduling in the PRC.

2.2.3 Achievements of energy-efficient power generation scheduling in Guangdong

In 2009, the power supply was sufficient in the first half year, and getting insufficient in the latter half year, this situation lasted the whole year in 2010. Through the implementation of energy-efficient power generation scheduling in Guangzhou Province, in 2009, in total 901,000 tons of coal equivalent was saved in coal-fired power plants in Guangdong province, accounting for 1.1% of coal consumption for power generation in the province, reduced 1.9822 million tons of CO_2 emission, 14,300 tons of SO_2 emission, with average desulfurization efficiency of 94.26% for coal-fired generation units.

From January to April 2010, cumulative savings of coal equivalent is 130,000 tons, reducing 286,000 tons of CO_2 emissions, 2,100 tons of SO_2 emissions, with average desulfurization rate of 93.74% for coal-fired generation units.

The energy-efficient power generation scheduling for different power sources was successfully carried out in Guangdong, including coal, oil, gas, water, nuclear, stored energy and other types. Heat load online monitoring system covering all provincial thermal power plants was established. And dispatching for 29 generation units from 12 plants was completed, to expand the coverage of energy generation dispatching.

2.2.4 Existing problems and development trends of energy-efficient power generation scheduling in Guangdong

Main problems of energy-efficient power generation scheduling in pilot areas in Guangzhou are as below:

(1) Insufficient power supply

When the power supply is sufficient, a positive impact on energy conservation can be achieved by sorting power generation units in accordance with the principles of energy-efficient power generation scheduling. However, when the power supply is insufficient, all generation units would get involved to power generation, or even subsidize power generation, which would cause negative impacts on energy conservation. Power supply in Guangdong is always insufficient, which seriously affected the implementation of energy-efficient power generation scheduling.

(2) Imperfect of grid structure

During the actual operation period, power transmission capacity is limited due to imperfect power structure, resulting in less power generation of large generation units. Thus it would affect energy-efficient power generation scheduling results.

(3) Thermal power generation units by "point to point" power transmission are not included into the sorting table

After the implementation of energy-efficient power generation scheduling, thermal power generation units using "point to point" power transmission are still got involved into the "west-to-east power transmission" program, but not included into the ranking list. According to the current compensation approach of backup peak shaving and frequency regulation, those generation units should be responsible for compensation of ancillary units. But it is difficult to carry out.

(4) The current economic compensation approach could not cover the cost of ancillary units

The current economic compensation approach could not cover the cost of generation units with task of peak shaving, frequency regulation and backup service.

(5) Relationship between energy conservation and reducing power generation cost

Contradiction exists between energy conservation and reducing power generation cost. Such as, in addition to hydropower, generation cost of renewable energy generation is more than coal-fired power generation. Cost of natural gas is relatively high. Giving priority to these two types of power generating unit may substantial increase the cost of electricity purchasing. For a long term, all factors should be fully considered, to deal with the relationship between energy conservation and reducing of power generation cost.

The future work of energy-efficient power generation in Guangdong Province includes the following aspects: Firstly, to improve technical supporting system of energy-efficient power generation scheduling, to enhance effects of energy-efficient power generation scheduling, to accelerate the establishment of coal consumption real-time research and coal consumption online monitoring system, gradually achieve the transition from ranking with planned coal consumption to ranking with real-time coal consumption; secondly, to strengthen the management of grid-connected power plants both in local place and corporate, to achieve unified dispatching of all power generation units; thirdly, to complete the study of energy-efficient power generation scheduling mode.

2.3 Implementation situation of energy-efficient power generation scheduling of China Southern Power Grid

2.3.1 Basic situation of China Southern Power Grid

(1) Current situation of power source

By the end of 2009, installed capacity of China Southern Power Grid reached to 135013 MW, the installed capacity for Guangdong, Guangxi, Yunnan, Guizhou and Hainan are 53327MW, 13188 MW, 19893MW, 18128 MW and 3427MW respectively. The power generation units are various, of which clean energy accounted for 37.7%, including 32.4% of hydropower, 2.9% of nuclear power, 2.0% of pumped storage units; and fossil energy based units accounted for 62.3%, including 53.0% of coal-fired units, 9.3% of oil-fired and gas-fired units.

The ratio of integrated hydropower generation units of China Southern Power Grid reaches to 32.4%. The ratio of both Guangxi and Yunnan is over 50%. As on-grid hydropower only accounts for a small proportion, the implementation of energy-efficient power generation scheduling has significant meaning to fully use of hydro resource. The ratio of integrated coal-fired generation units reaches to 53.0%. The implementation of new scheduling rule has significant benefit to reduction in coal consumption and flue gas emission.

(2) Current situation of power grid

Southern Power Grid, covering Guangdong, Guangxi, Yunnan, Guizhou and Hainan provinces, has 13 major power transmission lines from west to east, including "five DC, eight AC", AC and DC combined operation, long-distance, high-capacity, high voltage transmission. The maximum transmission capacity of electric power from West to East reaches over 23 million kWh.

By the end of 2009, in total, China Southern Power Grid has constructed 77 transformer substations with voltage of 500kV, and 145 voltage convertors with voltage of 500kW, 500kV power transmission capacity of 118000MVA, 269 lines with voltage of 500kV; 540 transformer substation with voltage of 220kV, 1147 voltage convertors with capacity of 188622MVA and 1479 lines with voltage of 220kV.

In 2009, the highest unified dispatched load of China Southern Power Grid reached to 95902MW. The whole dispatched power reached to 585.92 billion kWh, of which Guangdong, Guangxi, Yunnan, Guizhou, Hainan accounted for 60.1%, 11.9%, 12.5%, 12.6% and 2% respectively. In 2009, the highest power Guangdong province received was 20700MW from "west-to-east power transmission" program, reached to 95902MW for the whole year.

(3) Power dispatch situation

Dispatching department in Southern Power Grid is divided into four levels: overall dispatching of Southern Power Grid, provincial (regional) level, regional (city, state) level, county (county-level cities). Major generation units are dispatched by the general office of

Southern Power Grid and five provincial agencies, only few units are dispatched by local dispatching agency, normally hydropower.

The planning of energy-efficient power generation scheduling in China Southern Power Grid is prepared by CSG dispatching center and provincial dispatching center. Which means the CSG dispatching center is responsible for preparation of optimization of "west-to-east power transmission" program, and the provincial dispatching center is responsible for the preparation of optimization of provincial generation units. The method and procedure of preparation of energy-efficient power generation scheduling in CSG is as below:

According to "west-to-east power transmission" agreement and power transmission/accepting contracts reached among provincial governments, the CSG prepares the primary planning of provincial power transmission and receiving planning and sends to five provincial dispatch centers in accordance with water condition of direct dispatched hydropower plants.

Five provincial dispatch centers prepare provincial power generation curve in the next day based on units combination program and power load demand forecasting for next day, according to energy-efficient power generation scheduling rule; and submit the CSG general dispatch agency. If the province with rich hydropower resource can not receive related quotas, application of adjustment in power transmission planning and related description of actual situation should be submitted to the CSG.

If there is no application for adjustment of provincial power transmission planning, the CSG general dispatch center should conduct security check for provincial power grid. If necessary, appropriate adjustment should be made. The power transmission and receiving plan, as well as power generation plan of units can only be announced after the security check.

If any province submitted application of adjustment on inter-provincial power transmission planning curve, CSG general dispatch center should make suitable adjustment according to the principle of "give priority to inner provincial transmission, optimize inter-provincial dispatch". And security check should be conducted for CSG and provincial power grid. The power transmission and receiving plan, as well as power generation plan of units can only be announced after the security check.

2.3.2 Measures adopted by China Southern Power Grid in energy-efficient scheduling

Since July 21, 2010, the simulation operation of energy-efficient scheduling has been started in China Southern Power Grid. Before this, preparation was conducted in Guangxi, Yunnan, Hainan provinces in terms of technical supporting system and others. National Energy Administration issued "Implementation Program of EEPGS in Southern Power Grid " in December 2010, mainly clarified the guiding ideology, objectives, working principles, institutional organization, division of responsibilities, implementation body, work procedure, leading group and staff namelist. Based on the implementation program, according to overall requirement of "provincial sorting, regional optimization", the provincial or regional government organize related departments, authorities, power plants to carry out energy-efficient power generation scheduling work; the National Energy Administration organized related national departments, SERC, Southern Power Grid Company and related power generation corporate, to conduct regional optimization of dispatching work. On December 27, 2010, CSG announced that full implementation of energy-efficient power generation scheduling in five provinces of this region.

Provincial power management authorities, power dispatch agencies and related power plants should establish working principles and requirements according to "Implementation Rules of EEPGS (trial)" and provincial implementation program of energy-efficient power generation scheduling, to carry out units sorting, load forecasting, power generation combination and load allocation. The related information of unit sorting, power generation combination program, load forecasting and load allocation result should be submitted to general dispatch office of CSG.

Generally, inter-provincial power generation quotas exchange program should apply rules in accordance with inter-provincial agreement and power transmission and receiving contract. If the hydropower cannot be fully received, the CSG should organize all provincial power grid companies to adjust the inter-provincial power generation plan in accordance with principle of "give priority to inner provincial transmission, optimize inter-provincial dispatch", to maximum use of renewable resource.

2.3.3 Achievements of energy-efficient scheduling in China Southern Power Grid

It has been three years since the implementation of energy-efficient power generation scheduling in China Southern Power Grid. Sound institutional system, policy system and technical supporting system has been established in Guizhou and Guangdong provinces. Related technical supporting system for energy-efficient power generation scheduling has been established and put into operation, to achieve balanced coordination and attract hydropower integration.

In 2008, 2009 and 2010 (by November), about 8.779 billion kWh, 3.398 billion kWh and 4.468 billion kWh of hydropower was attracted, respectively in 2008, 2009, 2010 (to November), by reducing coal consumption of 2.801 million tons, 1.07 million tons and 1.385 million tons respectively. In total, 6.33 million tons of coal equivalent has been saved in the past three years.

In 2008, 2009 and 2010 (by November), CO_2 emission has been reduced by 13.93 million tons, as well as 120,000 tons of SO_2 with reduction in coal consumption. Through strengthening on desulfurization online monitoring, the desulfurization effects have been guaranteed, with average desulfurization rate of 94% in Guizhou and Guangdong. In total, about 4.786 million tons of SO_2 has been reduced.

2.4 Effect analysis of energy-efficient power generation scheduling in Jiangsu province

2.4.1 Basic situation of Jiangsu Power Grid

By December 31, 2009, the total installed capacity for unified dispatching in Jiangsu Power Grid reached to 51.29 billion kW. The number of units with capacity of 600,000 kW and above reached to 29, installation capacity of 20.09 million kW, accounting for 39.17%; units with capacity of 300,000~600,000kW reached to 63, with installed capacity of 21.244 billion kW, accounting for 41.42%; units with capacity of below 300,000kW reached to 128, with installed capacity of 9.995 billion kW, accounting for 19.41%. 96 network lines of 500kV were constructed with total length of 7614km. Power grid of 220kV and below will be operated by region.

2.4.2 Measures adopted for energy-efficient power generation scheduling in Jiangsu

On September 1, 2008, Jiangsu Grid launched the simulation operation of energy-efficient power generation scheduling. Jiangsu Province has developed "Implementation Rules of EEPGS in Jiangsu Province (Trial)", "Pilot Work Program of EEPGS in Jiangsu Province", " Economic Compensation Approach for EEPGS in Jiangsu Province (Trial)" and "Sorting Table of Generation Units in Jiangsu Province" etc.. Power trading is conducted. And technology supporting platform of energy-efficient power generation scheduling is established, as well as heat online monitoring system, desulfurization and smoke monitoring systems, energy consumption monitoring system and daily generation planning system and other technical support system

(1) Implementation rules of energy-efficient power generation scheduling in Jiangsu Province (Trial)

The "Implementation Rules of EEPGS in Jiangsu Province (Trial)" clearly defined specific implementation approaches of energy-efficient power generation scheduling in Jiangsu province. The principles of ranking are described as below: 1) unadjustable wind, solar, oceanic, hydro and other renewable energy generators; 2) adjustable hydro, biomass, geothermal, and other renewable energy generators, as well as solid waste-fired units which meet environmental requirements; 3) nuclear power generation units; 4) coal-fired cogeneration units, generators which use waste heat, residual gas, residual pressure, coal gangue , washed coal, coal bed methane as the power resource; 5) natural gas, coal gasification based; 6) other coal-fired generation units, including cogeneration without heat load; 7) oil-based and oil product-based generation units; 8) pumped storage units are excluded in the sorting table, which the generating quotas would be allocated by provincial power dispatching agency according to grid demands.

It requires power plants to install desulfurization and flue gas real-time monitoring systems and information management system, information management system of thermal power generation units in Jiangsu, and information management system of integrated resource utilization in Jiangsu. Waste based power generation units which cannot meet requirements of environmental protection and coal-fired units without installation of monitoring equipments or with abnormal operational monitoring equipments should be allocated any power generation quotas. Power plants submit actual parameters of operational units related to energy-efficient power generation scheduling and design parameters of units to be used in the next year by September 20 to provincial Economic and Trade Commission, Nanjing Electricity Regulatory Office, provincial power dispatch authority. By every November 20, the provincial Economic and Trade Commission prepares the sorting table for next year with consideration of unit type, energy consumption level of thermal power units and environmental factors. And the sorting table will be assigned to provincial power dispatch agency to implement and public release. The provincial power dispatch agency prepares units combination program for the next before 12:00 according to annual sorting table and the maintenance and commissioning situation of units.

The provincial Economic and Trade Commission conducts annual and monthly power load demand forecasting and release to the public through power regulatory website and "openness, fairness and impartiality" dispatching website. The major work of provincial power dispatching agency is daily load forecasting and management to balance power generation and power utilization. The provincial Economic and Trade Commission prepares provincial annual and monthly power generation combination program according to annual load forecasting, unit maintenance plan and actual operation situation of units. The provincial dispatch agency prepares provincial daily power generation combination program according to daily load forecasting and sorting table. Self-supply units will be involved into the sorting as well as power generation. In terms of maintenance, backup and security check, all grid-connected units must comply unified dispatching by power grid and participate peak shaving, frequency regulation and voltage adjustment, as well as implementation of daily dispatching plan issued by provincial dispatching agency.

(2) Pilot work program of energy-efficient power generation scheduling in Jiangsu Province

The "Pilot Work Program of EEPGS in Jiangsu Province" proposed the guiding ideology, objectives and working principles of pilot project in Jiangsu Province, clarified pilot institutional organization, and the main responsibilities, including, preparation of "Implementation Rules of EEPGS in Jiangsu Province (Trial)", preparation of units ranking list. It also clarified the compensation mechanism to unqualified units which has been decommissioned. Provincial Power Company is responsible for establishment and acceptance check for related technical supporting systems. Power dispatching department is responsible for security check. Flue gas online monitoring system should be installed for coal-fired generation units; heat load real-time monitoring system should be installed for CHP units; resource utilization real-time monitoring system should be installed for units with integrated resources. Coal consumption real-time monitoring and information management system should be developed. Thus, the sorting table of generation units can be prepared based on real-time coal consumption data.

(3) Economic compensation approach for energy-efficient power generation scheduling in Jiangsu Province (Trial)

The "Economic Compensation Approach for EEPGS in Jiangsu Province (Trial)" clearly defined specific measures of compensation for decommissioned units, compensation for energy-efficient power generation scheduling, economic compensation clearing, return of resource integrated utilization and heat supplying units. The objective of compensation is unit involved into annual power generation combination program and actual power generation amount is less than annual power generation baseline. As for coal-fired units, based on the sorting table and load allocation principle, power plant with high efficiency and high quotas should provide reasonable compensation for power plant with high energy consumption and less quotas, to ensure power generation during the peak period when the power grid needed, as well as demands for security backup and frequency regulation. Power generation quotas of inter-regional/inter-provincial bilateral transaction should complete clearing according to agreed electricity tariff. The allocated quotas for decommissioned units should transfer to coal-fired units with capacity of 300,000 kW. Compensation time for decommissioned units should be in line with "Suggestions Regarding to Speeding up Decommission of Small Thermal Power Generating Units" (State Council No. [2007]2).

Provincial Economic and Trade Commission determines annual power generation quotas baseline for generation units according to average social cost of coal-fired units and provincial power generation balance situation, and power generation quotas for the next month according to balance situation of monthly power generation and power utilization. The power generation baseline quotas apply province-approved electricity tariff. The unit utilization hour is determined by its efficiency. Power plant with high efficiency which the actual power generation amount is higher than the baseline quotas should provide compensation to power plant with high energy consumption level which the actual generation amount is less than the baseline quotas. Economic compensation applies principle of limited cost compensation, forms of transaction compensation and sharing compensation. The transaction compensation is implemented through monthly negotiation or market direct.

(4) Alternative power generation (power generation right trading)

The "alternative power generation" refers to replacement of inefficient small generation units with large efficient units, to optimize energy resources, save coal consumption, reduce pollutant emissions to achieve the fundamental goal of energy conservation.

1) Phase I

The first phase of "alternative power generation" happens in inner power plants. Optimization of units operation within power plant should be encouraged to decommission small unit with high energy consumption.

2) Phase II

The second phase of "alternative power generation" is market-oriented transaction power generation right. Since August 9, 2006, Jiangsu provincial Economic and Trade Commission and provincial power plant jointly issued "Suggestions Regarding to Optimization Allocation of Resources through Implementation of EEPGS in Jiangsu province" and related implementation specification issued by provincial power plants.

The scope of "alternative power generation" enlarged to provincial inner power generation group and between power plants. The alternative power generation volume increased to 200,000 kW from 135,000 kW. Alternative power generation is achieved through two approaches, one is negotiation, and the other one is market-matching. It gradually develops to power generation right trading with provincial dispatched units.

3) Phase III

The third phase of "alternative power generation" aims to energy conservation. In 2007, the Jiangsu Provincial Economic and Trade Commission and Nanjing Electricity Regulatory Office jointly issued "Guidelines Regarding to Implementation of Alternative Power Generation to Achieve Optimization Allocation of Resource in Jiangsu Power Grid in 2007".

A series of reforms have been carried out for trading rules. The old principle of "price priority, efficiency priority, time priority" was changed to "efficiency priority, price priority, time priority". It means that small units will trade their quotas to units with capacity of 600,000kW, then units with capacity of 300,000kW.

2.4.3 Achievements of energy-efficient power generation scheduling in Jiangsu

In 2008, in total, 39 power plants in Jiangsu Province are involved in alternative power generation, of which 15 plants were replaced, and 24 plants are alternatives. A total of 6.248 billion kWh is generated through power generation right trading, saving 300,000 tons coal equivalent, reducing 6000 tons of SO_2 emissions. About 3.01 billion kWh is from inter-plants replacement, 1.76 billion kWh from contracts trading, 14.3 billion kWh from market transaction. The average load of large units increased from 82.2% to 85.7%, and average load of replaced small units decreased from 85.9% to 83.5%. Average utilization hour of large units (600,000 kWh) increased to 5500~5700, with average increasing rate of 250 hours.

2.4.4 Existing problems of energy-efficient power generation scheduling in Jiangsu

Main problems of energy-efficient power generation scheduling in Jiangsu province can be summarized as below:

(1) The difference of power generation quota among large, small, decommissioned generation units is small. Annual quota allocated to small units is gradually decreasing. Besides, based on annual demand prediction, some discounted quota is reserved, mainly to meet demands of major users and users from other regions, which apparently will impact annual quota of power generation.

(2) After the implementation of energy-efficient power generation scheduling, it is getting harder to balance the interests between power generation companies and power grid companies. Due to some thermal power generation units are gradually replaced by renewable resource-based power generation units, compensation should be paid for some thermal power generation units. However, since agreement is hard to reach, reasonable distribution of benefits and compensation methods are not established yet.

(3) Energy-efficient power generation scheduling considered grid security constraints, but not fully considered the transmission loss.

2.5 Effect analysis of energy-efficient power generation scheduling in Sichuan

2.5.1 Basic situation of Sichuan Power Grid

Water resource is rich in Sichuan province resulting in big proportion of hydropower. By the end of 2008, the total installed capacity in Sichuan province reached to 26.83 billion kW, of which 15.46 million kW is hydropower units and 11.37 million kW of thermal power units. With further completion of a number of large hydropower projects, the proportion of installed hydropower capacity will greatly increase. As for the actual power generation, hydropower should be generated as much as they can during the flow season, while the thermal power units should be used at a minimum amount. And the thermal unit is essential during the dry season in the summer to ensure the safe and stable operation and meet the normal power demand.

The operation of Sichuan Power Grid is different from other provincial power grid, mainly characterized as:

1) Poor regulation capacity of reservoirs. Only six of grid-connected hydropower plants in Sichuan power grid have seasonal regulation function, such as Ertan and baozhushi. Other hydropower plants are based on daily/weekly regulating reservoirs or runoff regulation reservoirs which cannot fully use of water resource during the flow reason.

2) Poor peak shaving capacity of thermal power generating units. Due to poor coal quality and other reasons, the peak shaving capacity of thermal power generation units greatly decreased, resulting in peak shaving pressure of hydropower units during flow reason. In order to meet peak load demand, more thermal power units have to be allocated. In this case, certain power generated by hydropower units would be wasted.

3) There is big difference of power load between thermal power and hydropower

during different season and different time.

4) The water amount of each watershed is heavily affected by the climate. The runoff situation of each hydropower plant is uncertain. Big change of power load cause difficulties for power scheduling.

2.5.2 Measures adopted for energy-efficient power generation scheduling in Sichuan

On December 3 2008, Sichuan province launched energy-efficient power generation scheduling pilots.

(1) Overall measures of EEPGS in Sichuan

Two main stages are divided to implement the new scheduling rule:

The first stage (starting from the national approval), is to carry out pilot project of energy-efficient power generation scheduling for grid-connected power plants. Main work includes preparation of ranking list for generation units and generator combination plan. According to the implementation rules, it clarified compensation approaches for peak shaving, frequency regulation, maintenance, reserve, and compensation for decommissioned units. Thermal power plants gradually install flue gas online monitoring facilities and accessed to the network, report coal consumption parameters, install real-time heat load monitoring system for the heat load cogeneration units, and complete the technical preparation and safety check.

The second phase (starting 3 months after the pilot launched), is period of full implementation of pilot work, including two major parts. Firstly, in addition to grid-connected power plants, coordination should be made for non-connected power plants to get them involved into energy-efficient power generation scheduling. Economic commission at city level combined with related department should be responsible to preparation of ranking list of generation units and unit combination program to submit to provincial government for approval, afterwards, to be implemented by the provincial Power Company with consideration of coal consumption, security and resource optimal allocation. Secondly, local independent energy-efficient power generation scheduling work should be launched. Local economic and trade commission together with related department should be responsible for preparation of implementation suggestions for energy-efficient power generation scheduling, ranking list of generators, power generation combination program, as well as implementation of above works after approval.

At the policy level, units sorting table of main grid in the first quarter of 2009 was prepared, as well as implementation program of power compensation due to commissioning of small thermal power generation units in 2008. At the technical level, Sichuan Power Company completed the development of "Technical Supporting System of EEPGS" based on intelligent scheduling techniques. The "Coal Monitoring & Warning Information System" was applied by Chengdu Electricity Regulatory Office. And coal-fired power plants with capacity of 100,000kW in Sichuan province can directly input their coal monitoring and warning information from this system. This system improved the efficiency of coal monitoring and forecasting and warning of power grid operation.

(2) Implementation program of power compensation for decommissioned small thermal units in Sichuan in 2008

The "Notice of Implementation Program of Power Compensation for Decommissioned Small Thermal Units in Sichuan" is issued by General Offices of Sichuan People's Government in 2008 (Sichuan General Office No. 〔2008〕183) clearly clarified the compensation implementation details.

1) The range of involved hydropower plants, power quotas and implementation approach. Hydropower plants have to put into operation before October 2007, except those not put into operation after "5.12" earthquake. The compensated power quota is evaluated as 7% of actual on-grid power volume during flow season in 2007. For those plants who closed over 1 month after "5.12" earthquake but recovered and grid-connected, the compensation power quotas is half decreased. Detailed implementation approach and compensation amount should in accordance with Sichuan General Office No. 〔2008〕183 document.

2) Contract of replacement of thermal power and hydropower. Replacement contract signed between small thermal power plants and related hydropower plants should apply Sichuan General Office No. 〔2008〕183 document. The implementation of the contract should be completed through transaction platform of Sichuan Provincial Power Company.

3) Sichuan Provincial Power Company is responsible for detailed implementation work, strictly complying the principle and pricing according to Sichuan General Office No. 〔2008〕183 document, to ensure that compensation to decommissioned small thermal units can be completed before the flow reason.

2.5.3   Achievements of energy-efficient power generation scheduling in Sichuan

In 2008, during the high flow period, disposable water was decreased by 1.16 billion kWh compared to same period in 2007, saved coal of 813,000 tons and reduced 15,600 tons of SO_2 emissions.

Compared with provinces dominated by coal-fired power generation, the coal saving effect is more significant in Sichuan due to its rich hydropower resource. In 2008, through power generation right trading between hydro-thermal power plants, hydropower generation increased by 3.331 billion kWh, equivalent to reducing coal consumption of 1.118 million tons. Optimal scheduling of thermal power declined coal consumption of 7 grams per kilowatt-hour. In the first five months of 2009, utilization of generation units with capacity of 600 MW reached to 2255 hours, increasing of 864 hours than last year; utilization of generation units with capacity of 300 MW reached to 1286 hours, increasing of 91 hours than last year. On the contrary, utilization of generation units with capacity of 200 MW reached to 580 hours, decreasing of 338 hours than last year; utilization of generation units with capacity less than 200 MW reached to 1458 hours, decreasing of 495 hours than last year. Coal consumption for thermal power plants declined by 4.5 grams compared with previous year.

2.5.4 Existing problems of energy-efficient power generation scheduling in Sichuan

In Sichuan Province, due to the hydropower generating unit contains a great proportion; the existing problems of implementation the energy-efficient power generation scheduling are listed as below.

(1) Different load characteristics during the high/low flow period, results in different dispatching approach between thermal power and hydropower generation

The power supply capacity is heavily affected by seasonal water runoff. This means surplus electricity during high flow period and insufficient power during dry season. Therefore, the actual dispatching cannot be strictly conducted in line with the ranking list. Flexible arrangement should be prepared to ensure stable power supply.

(2) Grid security constraints affect the EEPGS ranking result of generation units

Due to weak connection between Sichuan Power Grid and the State Grid, network frame with capacity of 500 kV and 220 kV in most regions remains electromagnetic loop. The network still has many weak and dangerous points. The current units sorting method and load allocation result will result in overload or heavy load of some cross-section of transmission lines and low regional voltage, in following aspects: ① some thermal power plants can not sorted according to energy-efficient power generation scheduling rule due to grid security constraints; ② overload of transmission lines for hydropower; ③ operation of new equipments would impact current network distribution; ④ security constraints of power grid in some areas at specific time would impact units sorting of hydropower plants.

(3) Large-scale maintenance or special operation results in negative impacts on

scheduling implementation

During network improvement and development period, large-scale maintenance of the network or special operation mode, must impact the generating capacity of related generators, further affect dispatching effects.

(4) Impacts of EEPGS on "openness, fairness and impartiality" dispatching

The ranking list of generation units might adjust the power load of power plants, which would impact the implementation of "openness, fairness and impartiality" dispatching rule.

(5) Impacts of EEPGS on dispatching management

According to requirements of the implementation of energy-efficient power generation scheduling, a large number of local power plants would be included in the unified dispatching management. For those local power plants, uneven quality of personnel, backward equipment and technology, low level of automation and less self-consciousness of implementing dispatch instructions would result in some management problems.

(6) Impacts of internal factors of power plants on EEPGS results

Different historical conditions, composition, institutional organization, different management, and other reasons result in big different of operation conditions of power generation units. It also impacts dispatching management, including peak shaving capacity of generation units, coal supply situation, operation situation of generator and hydrological forecasting capabilities.

(7) Impacts of ancillary services of power plants on EEPGS

In 2009, the State Electricity Regulatory Commission issued a compensation provision of ancillary services from power plants, which would change the original ancillary service mode, eventually impact the energy-efficient scheduling results. Ancillary services compensation, such as one-time frequency regulation and on/off peak shaving, would also impact the energy-efficient scheduling results.

2.6 Effect analysis of energy-efficient power generation scheduling in Henan

2.6.1 Basic situation of Henan Power Grid

Henan Power Grid is the hub center of Center China, North China and West China power grids. The power grid network directs to north, west, central east and south of Henan province, formed four areas. All areas are connected by 500kV line. Electromagnetic round-loop with 500kV and 220kV exists within each area. Hydropower is concentrated in west Henan, mainly is Xiaolangdi Hydropower Plant with frequency regulation function. The characteristics of Henan Power grid is west-to-east and north-to-south power transmission. The first high-voltage demonstration project in the PRC: 1000kV Southeast of Shanxi-Nanyang-Jinmen across Henan province, which is effective in commercial operation.

By the end of December 2009, total installed capacity in Henan province reached to 46.798 million kW, of which hydropower reached to 3.65 billion kW accounting for 7.8%; thermal power reached to 42.94 million kW, accounting for 91.76%; new energy based power reached to 207,300 kW, accounting for 0.44%, including 48,800 kW of wind power and 156,000 kW of biomass based power.

Coal used for power generation in Henan is mainly from Henan itself, with supplementary of coal from Shanxi, Shaanxi and Mengxi. Coal transportation mainly depends on trail with complementary of highway. Between January and August 2009, the total capacity of decommissioned units reached to 3.48 million kW due to lack of coal. Fortunately, it did not cause great negative impacts on provincial power supply, but certain impacts caused in some areas.

In 2010, the annual surplus of electricity in Henan Power Grid is still sufficient after trading with outside power grid. The maximum surplus reached to 12 million kW. The backup rate of installed capacity reached to 54%. In 2010, the annual dispatching power volume will reach to 204.2 billion kWh. The dispatching power volume will reach to 203.05 billion kWh after exchange with outside Power Grid. The utilization hour of dispatching thermal units in 2010 will further decrease to 4521 hours.

It is estimated that the maximum load in Henan in 2011 will be 39.6 million kW, and the maximum load for unified dispatching will be 38.3 million kW. In 2010, thermal units of 1.198 million kW will be decommissioned. It is estimated that installed capacity in Henan will reach to 61.606 million kW by the end of 2011. The backup rate of installed capacity in the summer of 2011 will be 31%, and 40.7% in the winter. There will be surplus for the whole year. The utilization hour of unified dispatching thermal units will maintain to 4270 hours in 2011, decreasing by 251 hours compared to previous year.

2.6.2 Measures adopted for energy-efficient power generation scheduling in Henan

(1) Preparation work of EEPGS in Henan Province

The preparation of energy-efficient power generation scheduling in Henan Province includes establishment of provincial energy-efficient power generation scheduling organization, preparation of generator ranking list and combination program of generation units, improvement of technical supporting system, and compensation approaches of peak and frequency regulation and reserve units, network security check, improvement of grid network structure and information disclosure and monitoring.

1) Preparation of sorting table and unit combination program

The sorting table is prepared by Henan provincial Development and Reform Commission together with provincial environmental protection bureau, power company and related agencies, according to actual situation of Henan province and principle of energy conservation and environmental protection. According to actual operation situation of units, further optimization was conducted to sort heat cogeneration units and coal-fired units. The primary program of next-year energy-efficient power generation scheduling should be prepared and release to the public with full consideration of compensation for decommissioned units and backup units, security constraints and other issues.

2) Improvement of technical supporting system for energy-efficient power generation scheduling

The provincial environmental protection bureau is responsible for checking pollutant emission level of units, and promoting installation of flue gas online monitoring system according to requirements of energy-efficient power generation scheduling. Provincial power company promotes installation of auto power generation control, heat load online monitoring and flue gas online monitoring system and other technical supporting system.

3) Establishment of compensation approach for peak shaving, frequency regulation and backup units

Initiated by Zhengzhou Electricity Regulatory Office, together with provincial Development and Reform Commission, provincial power company and related agencies, "Compensation Approach for Peak Shaving, Frequency Regulation and Backup Units" is prepared. Main content of this approach including standard clarification of units participating in peak shaving, frequency regulation and backup (including cold reserve and heat reserve), and detailed compensation standard and methods.

4) Improvement of grid security check, stability and grid structure

The provincial power company is responsible for improving the check of grid security and stability, as well as grid structure, to ensure power security and stable power supply, and avoid impacts on the implementation energy-efficient power generation scheduling due to power grid restrict.

5) Information disclosure and regulation

Information disclosure is applied through whole procedure of energy-efficient power generation scheduling. Related methods of information disclosure and regulation are implemented according to regulations issued by National Electricity Regulatory Commission. All departments and agencies should provide demand information of energy-efficient power generation scheduling timely, accurately and completely according to related regulations.

(2) Compensation approach for peak and frequency regulation and reserve generation units in Henan province

The "Compensation Approach for Peak Shaving, Frequency Regulation and Backup Units in Henan province (Trial)" defines the scope of compensation and funding source; specific compensation approaches are established respectively, including compensation for automatic generation control, thermal power peak shaving, cold reserve, heat reserve. Besides, it also clarified the calculation and payment of compensation, and responsibilities of the various departments.

Sources of compensation funds include: 1) punishment capital from evaluation of provincial on-grid units; 2) fee charged for units which annual power generation hour is over than 5500 hours (monthly exceed 450 hours). Standard is 0.1 RMB/kWh, monthly charge, monthly billing and year-end counted. Monthly charging fee= rated capacity of unit×(monthly actual power generation hour−450) ×0.1 RMB/kWh. Annual charging fee =rated capacity of unit×(annual actual power generation hour−5500) ×0.1 RMB/kWh.

Compensation payment is monthly cleared and counted by the end of the year. If the same activity can apply different terms of compensation, the terms of highest compensation fee will be selected.

(3) Annual generation combination program of energy-efficient power generation scheduling in Henan province in 2009

The "Annual Generation Combination Program of EEPGS in Henan province in 2009" proposed combination program of the provincial power generation units and provincial power generation goal, as well as principles of annual combination of on-grid dispatching units and non-connected dispatching units.

As for unified dispatched units, the principles for identification of units combination program are: firstly, to encourage units with capacity of 300,000kW or above; the utilization hours of same type coal-fired units are similar: 3700 hours for units of 600,000kW, 3400 hours for units of 300,000kW and 3000 hours for units of 200,000kW; secondly, to ensure heat supply; the utilization hours of units which supply heat should be appropriately enhanced as: 3500 hours for units of 300,000kW, 3400 hours for units of 200,000kW or below; thirdly, utilization hours for hydropower units should be determined by average annual water amount; utilization hours for gas-fired units should be determined by contract gas supply amount in line with actual power generation amount in previous year; fourthly, to encourage clean production. Utilization hours of wastewater based units should increase by 50 hours; fifthly, preferential policy should be appropriately carried out for province-owned listed enterprises; utilization hours of foreign owned units should be allocated according contract; sixthly, considered with uncertainty of input, it is estimated that units which would be put into operation in the second half year will not be allocated by utilization hour; units which are put into operation in the first half year, the utilization hours should be same as units with same type. Self-supply units will not be involved into combination program; they are dispatched by principle of "self-production and self-consumption".

Annual combination program for non-unified dispatched units would not determined by power plant. Straw, wind, hydro, residual heat, residual gas, residual pressure, methane, coal bed based units can generate as much power as they can. Utilization hours of waste based units which meet requirements of environmental protection department should be determined according to total waste amount. The on-grid power price of above enterprises is determined by on-grid power clearing policy approved by central government. For other units, the annual utilization hours for provincial power generation would be 2500 hours.

To maintain continuity of compensation policy for decommissioned small thermal units, the compensated quotas, outside transmitted power amount should be included to annual power generation combination as incremental power amount. Based on voluntary principle, priority would be given to units ranked in the front of sorting table. Under the premise of ensuring safety and reliable power supply, alternative power generation should be encouraged; generation right of units with capacity of 200,000kW should be replaced by units with capacity of 600,000kW to reduce emission, save energy and improve economic benefits of power producers. Detailed transaction should be in line with related regulations.

Henan Power Grid developed the technical supporting system of EEPGS, online monitoring and data analysis system of EEPGS and information dissemination system, to provide safeguarding of the implementation of energy conservation dispatching at the technical level.

The technical supporting system of Henan energy-efficient power generation scheduling includes information collection, information exchange, dispatching management, real-time optimization coordination and control, assessment analysis and information dissemination. The system established impact assessment system of energy conservation dispatching, to evaluate the economic, social and environmental benefits generated from energy-efficient power generation scheduling, and provide information for policy maker.

Energy-efficient power generation scheduling online monitoring and data analysis system could conduct monitoring of real-time heat supplying situation of heat cogeneration units and scheduling range calculation, monitoring and evaluation of desulfurization equipment operation status of coal-fired units, monitoring of pollutants emission and sorting, as well as monitoring of real-time energy consumption.

On 27 June 2009, the information dissemination system of energy-efficient power generation scheduling in Henan was put into operation. This system proposed coordination mechanism and method of scheduling information integration, check, approval and release to ensure the utility and integration of the release system. Visualization method was applied for the first time in energy-efficient power generation scheduling information to resolve display problem of large amount data and multi-dimensional information. Security access control strategy was proposed aiming to different business, information and users, and a three-dimensional and multi-layer security system was established. Besides, the co-release mechanism of energy-efficient power generation scheduling was proposed to ensure the consistency of energy-efficient power generation scheduling information dissemination from different level of dispatching agencies.

2.6.3 Energy conservation and emission reduction results of Henan provincial power grid

1. Differential power generation quotas program

Under the condition of security constraits, Henan Power Grid ensures that high efficient units are allocated more power generation quotas according to principle of "high efficient and low energy consumption". Units which actual utilization rate is less than the limit standard of utilization rate, can trade their quotas and obtain financial compensation to maintain continuous power generation and reliable power supply within the province. Units which actual utilization rate exceeds the limit standard of utilization rate, can increase their quotas through bilateral negotiation or voluntary purchase, in accordance with the order of sorting table. Clearing of on-grid power is carried out by limit standard of power quotas, power quotas of decommissioned units, differential quotas with replaced units. If there is still surplus quotas, clearing is carried out as outside power transmission or market development.

According to above principles, power generation hour of different units under differential quotas index in 2009 was: 3700 hours for units with capacity of 600,000kW; 3400 hours for units with capacity of 300,000kW; 3000 hours for units with capacity of 200,000 kW; and 2500 hours for small thermal units; 400 hours for heat supply units; 50

hours increased for 600,000 units which use reclaimed water.

Power generation quotas for public-use coal-fired units in 2009 is as Table2-2.

Table 2-2   Energy-efficient scheduling index of coal-fired units in 2009

(Unit: 10,000 kW, 10,000 kWh, hour)

	Installed capacity	Total		Basic power amount		Voluntarily applied amount	
		Power generation volume	Utilization hours	Power generation volume	Utilization hours	Power generation volume	Utilization hours
Total	3450	17713766	5387	11594455	3526	6119311	1861
600,000	1464	7723955	5864	5016424	3809	2707531	2056
300,000~350,000	1119	5909169	5413	3691784	3382	2217385	2031
200,000	486	2323517	4466	1684947	3238	638571	1227
135,000 or below	381	1757125	4894	1201300	3346	555825	1548

It can be concluded from above table that, in 2009, utilization hour of units with capacity of 600,000kW increased 477 hours at average, about 970 hours more than units with capacity of 135,000kW; utilization hour of units with capacity of 300,000kW increased 26 hours at average, about 519 hours more than units with capacity of 200,000kW, which in line with the actual situation of Henan Power Grid and national policy energy-efficient power generation scheduling.

2. Power generation right trading

Through differential power generation quotas, decommission of small thermal units, power generation right trading and other market-oriented methods, large and high-efficient units obtain power generation quotas index, which greatly improved the utilization hours of unified dispatched units, reduced energy consumption and pollutant emission and eventually achieved energy conservation.

(1) Bigger proportion of power generation quotas for high-efficient large units

The proportion of power generation quotas for high-efficient units with capacity of 300,000 kW increased yearly, 67% in 2007 and 77% in 2009.

(2) Significantly increase of utilization hours of high-efficient large units

In 2009, the utilization hour of dispatched coal-fired units reached to 4721 hours, decreasing by 135 hours compared to 2008, details as below:

Units with capacity of 300,000kW or above: average utilization hour is 5005 hours, 284 hours more than average level, and 366 hours more than that of previous year; generated power increased by 16.6 billion kWh. Where: average utilization hour of units (including new units) with capacity of 600,000 kW reached to 5116 hours, increased by

324 hours than previous year; installed capacity increased by 3.12 million kW; generated power increased by 20.8 billion kWh. The average utilization hour of units (including new units) with capacity of 300,000 kW reached to 4784 hours, decreased by 630 hours compared to previous year; the installed capacity increased by 300,000kW; generated power decreased by 4.2 billion kWh.

Units with capacity less than 200,000kW: average utilization hour reached to 4055 hours, about 666 hours less than average level, maintaining the same as previous year; but the generated power decreased by 2.7 billion kWh compared to previous year.

In 2009, the utilization of units with capacity of 300,000 kW was 950 hours more than that of units with capacity of 200,000kW and below, which was in line with integrated power generation program issued by provincial government and national energy-efficient power generation scheduling.

Table 2-3 Comparison of EEPGS situation in 2008 and 2009
(Unit: 10,000kW, 10,000kWh, hour)

	Installed capacity			Generated power volume			Utilization hours		
	2009	2008	increase	2009	2008	increase	2009	2008	increase
Total (coal-fired unit)	3450	3199	251	1562	1423	139	4721	4856	−135
(1) 300,000 or above	2583	2241	342	1206	1040	166	5005	4639	366
Of which, 600,000	1464	1152	312	682	475	208	5116	4792	324
300,000	1119	1089	30	523	565	−42	4784	5414	−630
(2) 200,000 or below	867	958	−91	357	383	−27	4055	4001	54
Of which, 200,000	486	528	−42	205	220	−14	3946	4601	−655
135,000	381	430	−49	151	164	−12	4206	3911	295

As small thermal units are decommissioned, or units accessed to 220kV quit power generation due to itself problems, power source for 220kV is in a shortage. To ensure reliable power/heat supply, some units under 220kW with capacity of 300,000kW or below are allocated more hours. The average utilization hour of small thermal units in 2009 is 3326 hours, decreased by 215 hours compared to 2008.

3. Energy consumption index change

The proportion of large units has been increasing year by year, the management level has been improved, technical equipments have been gradually optimized, and coal consumption level has been declined yearly. Annual coal consumption index is shown in Table 2-4.

Table 2-4   Coal consumption for Henan power grid

(Unit: g/kWh, %)

item / year	PRC			Henan		
	Coal consumption for power generation	Coal consumption for power supply	Power utilization rate of power plants	Coal consumption for power generation	Coal consumption for power supply	Power utilization rate of power plants
2006	342	367	5.93	348	377	7.05
2007	332	356	5.83	334	361	7.11
2008	322	345	5.9	319	343	6.9
2009	319	342		312	334	6.36

It can be conculded from above table, coal consumption in 2006 and 2007 in Heinan is higher than national average value, but lower in 2008 and 2009. It indicates that implementation of EEPGS has positive impacts on reduction in coal consumption in Henan province.

2.7   Comparison analysis of energy-efficient scheduling in pilot areas

According to the "Energy-efficient Power Generation Scheduling Approach (Trial)" and the full simulation operation, the pilot provinces developed their own implementation dispatching approaches. In 2007, the capacity of decommissioned small thermal units in the PRC reached to 14.38 million kW, of which Jiangsu, Henan and Guangdong provinces ranked in top 5. The proportion of hydropower installed capacity in Sichuan province reached to 60%, it makes Sichuan province is the only province which the proportion exceed 50%. Due to different power source structure and different utilization situation of renewable energy, proposed pilot implementation program and coal conservation is different either. Among those pilot provinces, Guizhou and Guangdong provinces achieved significant effects of energy-efficient power generation scheduling.

2.7.1   Pilot workplan

Guizhou and Jiangsu Provinces proposed pilot workplan of energy-efficient power generation scheduling. Guangdong, Sichuan and Henan provinces proposed annual workplan of energy-efficient power generation scheduling, overall dispatching measures or preparation working program. The content of the pilot dispatching workplan is shown in Table 2-5.

Table 2-5 Workload comparison of pilot project in Guizhou and Jiangsu

Items		Guizhou	Jiangsu
Same points		Guiding principles, objectives, and institutional organization	
Different points	Working step	Proposed detailed plan for each stage, including task, working content, expected time and departments in charge	Proposed a loose plan and pilot working program, and task reporting plan of implementation details
	Other suggestions	Proposed to conduct energy policy and price policy research, bridging energy-efficient scheduling and power market reform, promote energy-efficient scheduling, and related departments in charge	
	Supporting measures and suggestions		Proposed the preparation of "Implementation details of energy-efficient power generation scheduling in Jiangsu (trial)", "Economic compensation approach for peak shaving, frequency regulation and reserve units in Jiangsu (trial)" and "Information dissemination method of energy-efficient power generation scheduling in Jiangsu", and relevant departments in charge
	Main conditions		Five conditions included policy level and technical level

Based on EEPGS simulation operation program, Guizhou Province established detailed pilot workplan, proposed detailed progress program, and work objectives, work content, timesheet and responsible authorities. It is helpful for the implementation and orderly progress of the new rule. Jiangsu Province proposed supporting measures and policy for energy-efficient power generation scheduling and defined the responsibilities of related departments, to provide policy supports.

2.7.2 Sorting table of generation units

Guizhou, Jiangsu and Sichuan provinces issued the sorting table of generation units, with their own characteristics in terms of generating unit type, and sorting principle.

Sorting table applied by Guizhou province divided the coal-fired generation units into four levels, including units above 600 MW, units of 300~600 MW, units of 200~300 MW and units of 100~200 MW. Units are sorted by those four levels based on coal consumption per unit (g/kWh). Priority is given to large units. It is possible that units with less capacity and less coal consumption rank in the front of the list.

Jiangsu province established the sorting table by planned coal consumption of generation units and performance parameters. After the establishment of real-time coal consumption monitoring system, the sorting table would be prepared in accordance with real-time value. The sorting table gives coal consumption for power generation, coal consumption for power supply, revised coal consumption for power supply (g/kWh). Unified sorting was prepared according revised coal consumption amount, from low to high.

The "Sorting Table of Generation Units in the First Quarter of 2009 in Sichuan province" divided generation units into thermal power generation units and hydropower generation units, giving priority to hydropower. Detailed ranking is as: 1) unadjustable hydropower units; 2) adjustable hydropower units; 3) biomass based units; 4) non-coal resources based integrated utilization units; 5) national demonstration projects; 6) heat cogeneration units; 7) coal gangue based integrated utilization units; 8) gas-fired units; 9) other coal-fired units. Adjustable hydropower units are ranked by its adjustment capacity from low to high. Heat cogeneration units are ranked its heat-power ratio from high to low. Coal gangue based units are ranked by coal consumption level from low to high. Gas-fired units are ranked by gas consumption rate from high to low. Other coal-fired units, unit without desulfurization equipments is ranked at latter position, in addition, others are ranked by coal consumption level from low to high.

The comparison of sorting tables in Guizhou, Jiangsu and Sichuan provinces is shown as Figure 2-3. The comparison of coal consumption for top-10 ranked units in Guizhou, Jiangsu and Sichuan provinces is shown in Table 2-6.

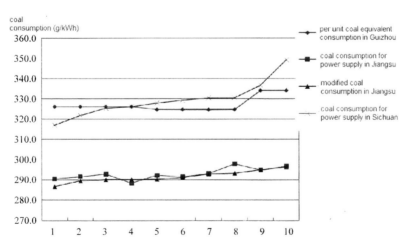

Figure 2-3   Coal consumption comparison for top ten coal-fired units in Guizhou, Jiangsu and Sichuan
Data source: (1) documents of EEPGS communication meeting of China Southern Power Grid;
(2) "Mid-term report of promoting EEPGS by foreign capitals";
(3) Sorting Table of Generation Units in the First Quarter of 2009 in Sichuan province.

Table 2-6 comparison of sorting table in Guizhou, Jiangsu and Sichuan

Items	Guizhou	Jiangsu	Sichuan
Unit type	Coal-fired generating unit	Coal-fired generating unit	Unified dispatching and distribution generation units of the main grid
Sorting rules	According to unit capacity, per unit coal consumption	Amended coal consumption	Unit type
Characteristics	Give priority to large units. Per unit coal consumption to prepare sorting table, but might not from low to high	coal-fired units are the dominant	Multiple power generating resources, hydropower installation capacity accounts for 60% of all sorted units
Items	Guizhou	Jiangsu	Sichuan
Coal consumption	High	low	High

It can be concluded that: in Sichuan province, hydropower resource is rich; the quantity of coal-fired generation units is small, with small installed capacity. The coal equivalent consumption per unit in Guizhou and coal consumption for power supply in Sichuan is higher than coal consumption for power supply and revised coal consumption in Jiangsu.

2.7.3 Economic compensation approach

Both Jiangsu and Henan provinces proposed trial economic compensation approach for energy-efficient power generation scheduling. Comparison results of two approaches can be found in Table 2-7.

Table 2-7 Comparison of economic compensation approach in Henan and Jiangsu

Items	Jiangsu	Henan
Applicable objective	Coal-fired generation units integrated to the grid in the province	Dispatching agency, grid integration company and grid company
Compensation objective	Units combination program listed to annual dispatching plan, units which actual generation capacity is lower the annual baseline	Economic compensation of AGC, deeply peak shaving, peak load shaving, heat back and cold back
Items	Jiangsu	Henan
Compensation clearing	economic compensation clearing	Compensation calculation method, including AGC, peak shaving of thermal power, compensation of heat and cold reserve
	Integrated utilization of unified resources and heat supply units clearing	Compensation fund source and account management

The objective and emphasis of those compensation approaches is different. The "Economic Compensation Approach for EEPGS in Jiangsu Province (Trial)" focuses on compensation to decommissioned units, energy-efficient power generation scheduling compensation, clearing of economic compensation, integrated utilization of dispatched resources, clearing of units supplying heat and other detailed operation approaches. The economic compensation approached applied in Henan province clarified the detailed compensation approaches for auto-generation control, peaking shaving of thermal power, cold and heat reserve.

2.7.4 Implementation specifications

Comparison of pilot implementation regulations of energy-efficient power generation scheduling in Guizhou and Jiangsu is shown in Table 2-8.

Table 2-8 Comparison of implementation specifications of energy-efficient power generation scheduling in Guizhou and Jiangsu

Items	Guizhou	Jiangsu
Same point	General principles	General principles
	Sorting of generation units	Dispatching sorting management
	Load forecasting and units combination	Load forecasting management
	Preparation of daily power generation plan and security check	units combination and load allocation
	Maintenance, peak shaving, frequency regulation and reserve	Maintenance, reserve and security check
	Appendix	Appendix
Different point	Interpretation of terms	Abnormal and emergency solution
	Emission inspection	Regulatory management
	Information dissemination	

(1) Management of generating unit sorting

Coal-fired generation units in Guizhou are sorted according to coal consumption, giving priority to energy-efficient units. Units with same coal consumption are sorted by pollutant commissions. The energy consumption value temporarily use designed value by manufacturing plant. But the real-time measured value would be used for sorting eventually. Appropriate adjustment should be made due to increase of real-time coal consumption due to environmental protection and water conservation reasons. The pollutant emission level should be in line with newest measured value by provincial environmental protection departments.

Jiangsu province defined that pumped storage units are excluded in sorting. If needed, the provincial dispatching agency will arrange the power generation according to the actual operation situation of power grid. Related power plants should install "Flue gas desulfurization real-time monitoring and information management system of coal-fired units in Jiangsu province", "Management information system of cogeneration units in Jiangsu province" and "Information management system of resources integrated utilization in Jiangsu province". Waste-based units which cannot meet requirements of environmental protection and coal-fired units without installation of monitoring equipments are not allowed to allocate generation quotas.

Guizhou province requires power plants submit parameters in every September, including the restriction level of reservoir during different period, changes of reservoir characteristics; integrated utilization requirements of hydropower plant; efficiency curves and water consumption rate curve of hydropower generation units; unit on/off time, minimum on/off interval time, load speed and safety operation parameters and environmental index approved by the Ministry of Environmental Protection.

(2) Compensation approach of peak shaving and frequency regulation

Guizhou province did not issue any economic compensation approach related to peak shaving and frequency regulation yet. The pilot implementation rule defined that compensation for generation units with ancillary services should be provided to ensure power quality and safety operation. Compensation details should be in accordance with "Management Approach of Scheduling Assessment for Power Plants in Guizhou"

The peak shaving of Guizhou Power Grid is mainly undertaken by hydropower generation units and coal-fired power generation units. The peaking shaving capacity of on-grid thermal unit should not be less than 50% of unit's rated capacity. The following factors are considered when decommission coal-fired generation units for peak shaving: i) should meet security constraints, without impact on continuous stable power supply; ii) generation units with bad performance of peak shaving and those cannot meet peak shaving capacity should be considered to give decommissioning priority; iii) units at back position in the ranking list should be considered for decommissioning.

Jiangsu province requires that all integrated generators must obey the unified power grid dispatching, to participate peak shaving, frequency regulation and voltage regulation, etc.

(3) Inspection of pollutants emissions

It is required to install flue gas online monitoring system for coal-fired generation units in Guizhou province, to accept the real-time dynamic monitoring from Guizhou provincial grid dispatching department. Besides, according to the requirements of "Technical standard of desulfurization information monitoring system for on-grid thermal power plants in Guizhou province", related desulfurization data should be provided to provincial grid dispatching department, being ensure of the accuracy, timeliness, and completeness of provided information. For those units, whose data has not been submitted due to equipment problem, should be considered unqualified in pollutant emission, and reflected in the ranking list of power generation units.

A qualified coal-fired units in desulfurization should be in line with following three indicators: ①desulfurization rate should not be less than 80%; ②the discharged SO_2 density should be less than 400 mg/m^3 or 1200 mg/m^3 according to different operation time; ③ the total desulfurization amount should be more than 70% of equipment designed desulfurization amount.

(4) Countermeasures for abnormal and emergency situation

Jiangsu Province declared that dispatching on-duty office can adjust combination of generation units and load allocation under abnormal or emergency situation. After abnormal or emergency called off, it should be gradually adjusted to new combined units in accordance with the ranking list.

Abnormal and emergency of power system could include: ① major accident of power generation/supply equipments or incident of power grid; ② grid frequency or voltage beyond the specified scope; ③ load of transmission and distribution equipment exceeds the specified value; ④ The power value of tie-line exceeds the stability limit; ⑤ due to weather and other reasons, actual load deviates from the expected load and it is difficult to adjustment; ⑥ renewable energy generation units can not fully access; ⑦ other emergencies threaten the safe operation of power grid.

2.7.5 Technical supporting system

All pilots have developed their own technical supporting system for energy-efficient power generation scheduling. (Table 2-9)

Table 2-9   Comparison of technical supporting system in five pilot provinces

Items	Guizhou	Guangdong	Jiangsu	Sichuan	Henan
Coal consumption online monitoring system	Yes	Yes	yes		yes
Desulfurization online monitoring system	Yes	Yes	yes		yes
Heat load monitoring system		Yes	yes		yes
Bus load forecasting		Yes		yes	
Security checking and correction	Yes	Yes		yes	
Dispatching generation system	Yes		yes		
Information dissemination system	Yes	Yes		yes	yes

All technical supporting systems of energy-efficient power generation scheduling in Guizhou, Guangdong, Jiangsu and Sichuan provinces have functions as on-line monitoring of coal consumption and desulfurization equipment operating. The MIS developed by Guizhou province integrated online coal consumption monitoring; UC / EDC power generation plan, information disclosure and other functions. Guangdong Province implemented compensation and support services and network operation assessment system; Sichuan generation is considered in the planning process the hydro-thermal coordination; Henan information disclosure system has better interactive features.

By the end of May 2010, comparison of pollutant online monitoring system between Guangdong and Guizhou is shown in Table 2-10.

Table 2-10   Comparison of pollutants online monitoring system for units in Guizhou and Guangdong

Items		Guizhou	Guangdong
Coal consumption online monitoring system	Trial operation time	April 2008	April 2010
	Number of dispatching units (unit)	56	-
	Number of units conducted monitoring (unit)	44	5
	Access rate (%)	79	-
Flue gas desulfurization online monitoring system	Trial operation time	July 2007	October 2008
	Number of units with desulfurization equipment (unit)	112	37
	Number of units with real-time monitoring function (unit)	112	37
	Access rate (%)	100	100

Data source: communication document of South China Grid energy-efficient scheduling pilot meeting.

Guizhou Province pioneered to develop coal consumption online monitoring system and flue gas desulfurization online monitoring system. The earliest applied system is online monitoring system for two pollutants. The installation rate of flue gas desulfurization online monitoring system in both provinces reached to 100% with well operation situation. In Guangdong, units with online desulfurization equipment reached to 112. In Guizhou, 12 generation units with capacity of 135 MW and 125 MW have been decommissioned; 44 units have complete installation of coal consumption online monitoring system, of which 40 units have been effective; the operation rate of coal consumption online monitoring system reached to 71%. Guizhou Province announced quarterly measured generated electricity and coal consumption for power supply, to prepare ranking list by real-time measured coal consumption. Measured coal consumption is more realistic, but requires large investment, with difficult of implementation. Guangdong Province prepares the ranking list by designed coal consumption. The measured coal consumption value is now gradually replacing the designed value. Sorted by designed coal consumption is simple to conduct, but the sorting result has slightly different with method by actual coal consumption. Compared to coal consumption rate for power supply, coal consumption rate for power generation is more likely to be convinced. Generating unit types in Guangdong province is various, almost covered all units defined in the implementation regulation of national energy-efficient power generation scheduling.

2.7.6 Differential power quotas

Both Sichuan and Henan provinces implement the differential power quotas program, with different standards.

Henan province established differential power quotas program, reflected in the annual electricity generation plan. The weight was established through power generation capacity, environmental protection and regional difference, to arrange generation hours of different generation units.

The principle of differential power generation plan for Sichuan grid is: thermal generation units are divided into four grades for capacity arrangement (100,000 kW, 200,000 kW, 300,000 kW and 600,000 kW), 100 hours difference between two grades. In the premise of completing the power plan, large units are allowed to replace small units for power generation. The settlement electricity tariff uses price applied by the transferor. The average utilization hour of circulating fluidized bed units should increase by 100 hours. The average utilization hour of hydropower plant which participate watershed compensation should increase by 100 hours as well.

2.7.7 Differential power pricing

Since 2007, both Sichuan and Guizhou provinces conducted the differential power pricing policy, to further strengthen the implementation of this policy. (Table 2-11)

Table 2-11   Comparison of differential power pricing

Items	Guizhou	Sichuan
Same point	Differential price of high energy-consumption and high pollution industry, desulfurization power price	
Different point	Residual heat and pressure grid price	Grid price of small thermal power plant
	CBM power price	punitive tariffs for over energy-consumption standard products

(1) Differential electricity pricing policy aimed to high energy consumption and high pollution industry

Since 2007, Guizhou province started differential power pricing policy for the following six categories characterized by high energy consumption and high pollution: ferro-alloys (including silicon, electrolytic manganese metal), steel, cement, calcium carbide, yellow phosphorus, lead and zinc. Meanwhile, three levels are divided for those industries into abandon, restricted abandon and restricted. Through this measure, in 2007, only 33 of 124 high energy consumption enterprises are still in operation in Guizhou province, and 91 of them have already decommissioned. In 200, the Guizhou Power Grid issued "Notice Regarding to Blackout Some Enterprises Implementing Differential Power Price". Surcharge power price due to differential power pricing should be conducted. For enterprises which are not pay surcharge power price according to differential power price policy, they will be blackout till they pay the surcharge amount. For enterprises which did not pay the listed power price, they will be regarded as self-decommissioned, and they will be permanently blackout, and power supply contract is regarded as completion. Since June 2010, the application of differential price policy is extended to 8 sectors, included electrolytic aluminum, ferroalloy, steel, calcium carbide, sodium hydroxide, cement, yellow phosphorus and zinc smelting sectors

The implementation of differential power pricing policy started in Sichuan in 2007, eight categories characterized by high energy consumption and high pollution are included, which are the electrolytic aluminum, calcium carbide, caustic soda, cement, steel, phosphorus, zinc smelting industries.

Since June 1 2010, tariff for restricted industry in Guizhou and Sichuan increased to CNY0.1 from CNY0.05/kWh; as for abandoned enterprises, it increased to CNY0.3 from CNY0.2/kWh.

(2) Desulfurization price

Dated by June 1 2010, desulfurization pricing policy is carried out in Guizhou province for qualified coal-fired generation units, with CNY0.017/ kWh. For those can not meet requirements of operation rate of desulfurization equipments, desulfurization price should be charged and punishment should be conducted.

By the end of May 2009, Sichuan province approved desulfurization pricing for 7 newly operational thermal power plants. Desulfurization pricing and desulfurization equipment operation monitoring mechanism was established to promote energy conservation and emission reduction.

(3) On-grid power price for residual-heat/residual-pressure power generation, coal bed methane power generation and small thermal power plants

Dated by June 1 2010, Guizhou province established preferential policy for energy efficient power generation projects, including pricing system for residual-heat/residual-pressure power generation, coal bed methane power generation and small size thermal power plants. Price for coal bed methane based power is CNY0.25/kWh more than on-grid desulfurization benchmark price in 2005.

By the end of May 2009, electricity price for 22 small size thermal power plants has been decreased in Sichuan, to promote decommissioning of small thermal units.

(4) Punitive tariffs for over energy-consumption products

Dated from June 1 2010, punitive tariff for over energy-consumption products is launched in Sichuan. Focusing on Iron and steel, ferro alloys, chemicals, cement and other "double high" industries, ranking list was proposed for those which has over energy-consumption. If the energy consumption is doubled more than the acceptable limitation, pricing system should be carried out as abandoned class; if it is within the standard limits, adopt the restricted corporate standards.

2.7.8 Power generation right trading

Trading of power generation right is conducted mainly in Jiangsu, Sichuan and Henan provinces. During the implementation of this policy, Sichuan adopted different measures from the other provinces, due to different energy source structure. (See Table 2-12).

Table 2-12 Comparison of power generation right trading measures in Jiangsu and Sichuan

Provinces	Jiangsu	Sichuan	Henan
Measures	Negotiation, market-matching, hydropower replaces thermal power	hydropower replaces thermal power	Aims to small units doesnot need to be decommissioned, and with capacity less than 200,000 kWh Dispatching units by negotiation and encouragement

Jiangsu Province conducted power generation right trading first, adopting "negotiation" and "market-matching" two alternative transaction methods. "Negotiation" is a free alternative of trading; "market-matching" is drawing on the principles of stock market transactions, through Jiangsu Power Grid trading center to complete the transaction. It clearly required unit with capacity of 200,000 or below must provide 50% of its power generation quotas to large units; the "negotiation" based power generation rights trading can only happened between small thermal units and units with capacity of 600,000kW. In 2008, in total 16.26 billion kWh of alternative power generation was completed; average coal consumption of thermal unit was 325 g/kWh, lower than national average level by 25 g/kWh, saving coal equivalent of 810,000 tons and coal cost of CNY 700. In July 2009, the power amount through power generation rights trading reached to 2.31 billion kWh.

In Sichuan province, the power generation rights trading happened through hydropower and thermal power exchange market. By March 2008, in total, power amount through power generation rights trading reached to 7.7 billion kWh. Generated hydropower increased by 7.7 billion kWh, 9 million m^3 of wastewater was reduced, 3 million tons of coal equivalent was saved resulting in SO_2 emission reduction of 120,000 tons and 2.1 million tons of dust and flue gas.

Henan province issued the "Interim Implementation Approach for Planned Power Generation Quotas Trading in Henan", to promote a province-wide implementation. From 2006 to September 2008, in total, 210 units in Henan participated power generation rights trading with total capacity of 3.504 million kW, saved coal equivalent of 1 million tons resulting in SO_2 emission reduction of 50,000 tons.

2.7.9 Program of "Large Substitutes for Small"

Henan province started the LSS program: firstly, small thermal power generation units which are not qualified to related policy should be decommissioned, and a list of decommissioning small thermal units with capacity of 2.4 million kW was prepared between 2005~2007; secondly, county-level dispatching for small thermal power plants has been adjusted to provincial dispatching, to strengthen power market management; and differential power generation quota has been conducted. Thirdly, power generation right trading has been carried out. Fourthly, the transformation from small thermal power plants to straw power generation is encouraged.

Guangdong province prepared "Implementation Program of Decommissioning Small Thermal Power Generation Units in Guangdong Province". It defined the decommissioning small thermal generators with capacity of 50,000 kW or below during the 11th "Five-Year Plan", to ensure the assignment of decommissioning small thermal units with total capacity of 9 million kW, or 10 million kW with efforts. According to the timeline, total capacity of decommissioned small thermal units in 2007 will reach to 2.20 million kW, and 3.85 million kW in 2008, 3.01 million kW in 2008. Guangdong province uses the pricing system to strengthen on-grid power price management for small thermal units, and to decrease on-grid power price of coal-fired small thermal unit to a level lower than provincial benchmark on-grid power price, without compensation. Since 2008, direct/indirect compensation for peak power generation of coal-fired units was no longer exists.

Guizhou province adopted the policy of giving priority to large thermal power generator. Small inefficient generation units are forced to be decommissioned. Large units with high capacity, high parameters, low energy-consumption and less pollutant emissions are encouraged.

2.7.10 Existing problems of energy-efficient power generation scheduling in pilot provinces

The main problems of energy-efficient power generation scheduling existed in pilot provinces include coordination between energy-efficient power generation scheduling and safety operation of power grid increasing of power supply cost, backwards economic compensation approach accuracy of technical supporting system, rational ranking of generation units and linkage between energy-efficient power generation scheduling and market mechanism.

(1) Coordination between energy-efficient power generation scheduling and safety operation of power grid

The energy-efficient power generation scheduling changed the original organizational structure and operation mode. It requires higher technology for power grid operation control. The new scheduling would cause negative impacts on safety operation of power grid. For instance, some units could not commission due to grid connection. Or some units should be involved into power generation even though they are not supposed to generate under the new rule in terms of security and stability of power grid. In some provinces with big proportion of hydropower, different scheduling programs should be carried out due to flow seasons to ensure the security of power grid. Besides, large-scale renewable energy would impact the grid safety. Therefore, the new scheduling rule should be further coordinated with safety operation of power grid.

(2) Increase of power supply cost by implementation of energy-efficient power generation scheduling

Decommission of small thermal generation units increased the investment of power grid, as well as cost of power supply. The program of "Large Substitutes Small" requires Grid Company to accelerate grid network construction, to ensure the power supply after decommissioning of small units. The power supply cost of Sichuan province remains the same as conventional scheduling; the power supply cost in Jiangsu and Henan provinces increased by CNY 1.128 billion and CNY 856 million respectively. It resulted in an average increasing of electricity price of CNY 0.0044/kWh, and CNY 0.0058/kWh. Take the Jiangsu province for example, the on-grid power price of new energy, such as wind power, nuclear power, pumped water storage and waste based power, ranges from CNY 0.455/kWh to CNY 0.90/kWh, much higher than that of coal-fired power; the electricity tariff in some areas becomes more and more higher due to surcharge of desulfurization price for coal-fired units. During the 11th Five-Year Plan, the total capacity of units which are installed desulfurization equipments reached to 171 million kW, calculated with compensation standard of CNY0.015/kWh for units with desulfurization equipments, the total compensation fee would reach to CNY 28.9 billion. However, in some places with high proportion of thermal power generating units, due to heavier assignment of installation of desulfurization equipments, the power purchasing cost is becoming much higher; for example in Beijing, Tianjin and Tangshan areas, the total capacity of units required installation of desulfurization equipments is 12.027 billion kW, to implement the desulfurization power price, the newly added power purchasing cost would be CNY 379 million. In Henan province, the approved capacity of units installed desulfurization equipments is 12.185 billion kW, of which 7.86 million kW is not involved into price adjustment program, another 6.245 million kW is not in operation yet, in total, about CNY 436 million will be increased to purchase electricity.

(3) Backwards economic compensation approach for energy-efficient power generation scheduling

How to reasonable balance stakeholders' interests is the essential problem need to be resolved. Under the new scheduling rule, all on-grid generation units should be involved into frequency regulation, peak shaving and backup. Economic compensation for those units due to reduction in power generation quotas or decommissioning should be prepared. An advanced compensation approach is important to the implementation of new scheduling rule. In the past, as all units are allocated same level of planned power generation quotas, as well as obligation of adjustment for power grid, the compensation was not considered.

Among five pilot provinces, Jiangsu province established "Economic Compensation Approach of EEPGS in Jiangsu Province (draft)". Henan province prepared "Peak Shaving, Frequency Regulation and Backup Compensation Approach of EEPGS in Henan Province (draft)".

(4) Accuracy of energy-efficient scheduling technical supporting systems

For the accuracy of coal consumption online monitoring, there is no standard for calculation method of coal consumption; the data used for the online system is manually input; except Guizhou province, the coal consumption online monitoring systems of other four pilot provinces are still under development or commissioning. Therefore, the technical supporting system needs further improvement.

(5) Rational ranking of generation units

Each ranking principle of generation units in pilot province has its own characteristic. Some generation units cannot be ranked by coal consumption data obtained from coal consumption online monitoring system due to backward technical supporting system. Some units are set of planned coal consumption during the construction period. Therefore, the quality of coal consumption online monitoring system would impact the generation units ranking.

(6) Linkage of energy-efficient power generation scheduling and market mechanism

Since the quota of scheduled generation unit is allocated in accordance with coal consumption.　It can only reflect the coal consumption level of the unit, instead of fixed investment cost, management cost, cost of coal transportation and other important factors. The energy-efficient power generation scheduling might reduce coal consumption of power generation, but increase comprehensive coal consumption, which would not fully meet the requirement of market mechanism.

(7) Function to emission reduction in Greenhouse Gas

The energy-efficient power generation scheduling emphasize on energy conservation, which is overall in line with emission reduction goal. During the dispatching arrangement, carbon dioxide emission of generation units should be optimized as quantitative parameter. However, the current energy-efficient scheduling cannot provide a specific method for that.

3. Management System of Energy-efficient Power Generation Scheduling

The primary basis of the implementation of energy-efficient power generation scheduling is construction of information platform with functions of data collection and sorting. In the power market, the information platform of energy-efficient power generation scheduling is decision-making supporting system of modernized power scheduling through power transaction in accordance with related regulations and management approaches of energy-efficient power generation scheduling.

3.1 Business framework of energy-efficient power generation scheduling

3.1.1 Business content of energy conservation power generation

The structure of power grid in the PRC is characterized is characterized as typical hierarchical partitioning. The implementation department of energy-efficient power generation scheduling is dispatching center of power grid, which is divided into several departments. The implementation of energy-efficient scheduling is realized through cooperation among all related departments from the dispatching center. The system platform of energy-efficient scheduling is constructed in the basis of electric power system and its automation, water resources and hydropower automation, energy and power engineering, computer information and network project. The business content of current energy-efficient power generation scheduling in the PRC includes eight aspects as below:

(1) Index calibration subsystem of thermal characteristics for thermal units involved into power transaction

Index calibration of thermal characteristics for thermal units involved into power transaction refers to boiler efficiency testing and turbine heat consumption based the principle of "openness, fairness and impartiality", and thermal characteristics of thermal units under different load through public heat testing. Here, we call it "series of certification coal consumption for units", which refers to coal consumption verified and recognized by a third party.

(2) Sorting table of local energy-efficient power generation scheduling prepared by local government according to annual power generation planning and coal consumption of units

Power management divisions of provincial development and reform commission prepare provincial annual power generation dispatching goal based on national annual power generation goal issued by NDRC and local economic development demands, as well as annual power purchasing plan of power grid companies, and prepare and backup sorting table through negotiation with power plants in accordance with the principle of "openness, fairness and impartiality". The monthly power generation plan of power plants should be issued by dispatching agency.

(3) Period division of energy-efficient power generation planning

Units involved into energy-efficient power generation scheduling should identify their on-grid power generation quotas by period. Therefore, 24 hours should be averagely divided by integer multiple number. Units are sorted according to their energy efficiency level. In Guizhou province, it divided into 96 periods, which means units are sorted every 15 minutes to track real-time change of energy efficiency of on-grid units.

(4) Issue annual energy-efficient power generation planning

At present, under the premise of ensuring safety power grid, we make every possible effort to ensure independent power plant to complete contractual power generation quotas. If the power generation planning changed due to market demand, adjustment of annual power generation planning should be made with the same proportion for all on-grid units in favor of clean energy. In this case, units of the same type should be allocated same utilization hours; utilization hours of high-efficient and low-pollution units should be much higher than that of low-efficient and high-pollution units; and power generation priority should be given to units with desulfurization equipment.

(5) Power grid issues load curve at each period according to annual energy-efficient power generation planning

Utilization hour of units at each period is determined according to coal consumption level. Every day is divided into several time intervals, units are sorted at each time period and related load curve can be formed according to on-grid power generation plans, to provide basic reference for dispatch officials to implement energy-efficient power generation scheduling.

(6) Hydro-thermal power complementarities

The energy-efficient power generation scheduling supports giving hydropower plants more power generation quotas to avoid water waste in power plants through actively promoting power generation right trading approaches like "hydropower substitutes thermal power" and LSS. Economic benefit of hydropower plants due to increased quotas can be used to compensate thermal power plants. During the flow season, cheaper electricity price can be adopted to sell more electricity to promote flow season power price. In some places, direct power supply to major users from independent hydropower plant can be conducted; the power grid company can charge a certain reasonable on-grid fee to promote fully utilization of hydropower resource. Besides, in addition to giving gird connection priority to wind power, geothermal power and other new energy based power, hydropower units can also be bundled with new energy based units at intervals. In this case, the preferential policies applied to renewable energy based power should be also applied to power generation amount of hydropower units using bundled grid-connection approach.

(7) Decommissioning small thermal power plant and its compensation

Decommission of small thermal power plants and power generation right transaction is a program that decommissioned small thermal power units are not arranged to connect the power grid. To ensure the smooth transition of small thermal power plants, the original allocated power generation quotas to small thermal power plants can be traded through power generating right transaction to compensate those small thermal power plants.

(8) Timely clearing

Clearing is timely settlement between power producer and power grid enterprise. The power buyer pays power purchasing fee to independent power producer. Timely clearing is in favor of capital flow of power producer. Transaction settlement adopts monthly pre-clearing, annual clearing to avoid impacting on power generation quotas. As for compensation for decommissioned small thermal power plants from LSS program, the power grid company finishes settlement with power producer first, then power producer provide compensation for small thermal units which are involved into power generation.

The difficult of construction of EEPGS platform is current power industrial structure adjustment—decommission of small thermal units. During system construction of EEPGS platform, we will have to face the integration of stored power generation assets, for example, small unit and unit with high coal consumption will be compensated according to original power generation planning; different types of EEPGS will be classified according to different compensation methods.

We should insist "openness, fairness and impartiality" principle, and increase the transparency of scheduling through information release platform. The EEPGS agency should release scheduling program for renewable energy, coal-fired units and scheduling results by month, quarter and year, including: power demand, power utilization rate, security operation parameter of power grid, on-grid power volume, price, power generation load rate, peak shaving and frequency regulation information and so on. The dispatch agency should consciously accept the supervision from government regulatory departments and community. Major issues should be reported promptly to power management department.

3.1.2 Business procedure of energy-efficient power generation scheduling

(1) Working procedure of technical supporting system of EEPGS

The energy-efficient power generation scheduling is a new procedure to reform power grid dispatching based on the current scheduling rule in accordance with "Notification Regarding to Publishing Pilot Workload and Implementation Rules of EEPGS" (NDRC Energy No.[2007]3523 document) issued by the NDRC and "Several Suggestions Regarding to Optimization Power Resource Allocation and Promoting Open and Fair Scheduling" (National Economic and Trade Commission Electric power No. [1999]1144 document). The procedure of energy-efficient scheduling includes four parts, as below:

i) to unify the third party certificate of real-time coal consumption of power generation units in different load;

ii) ranking of power generation units under energy-efficient scheduling rule;

iii) electricity sale according to the coal consumption (give priority to least coal consumption), power trading includes power generation quotas declaration, on-grid electricity price, efficiency ranking, units combination, safety checking, information dissemination and so on;

iv) daily information dissemination of trading data.

The main content of energy-efficient scheduling can be found in Figure 3-1.

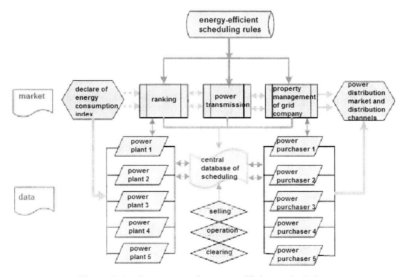

Figure 3-1   Procesure of energy-efficient scheduling

(2) Business framework under EEPGS rule

Reconstruction of the framework of energy-efficient power generation scheduling and procedure design is the milestone of Chinese power scheduling. Implementation of energy-efficient scheduling could promote the establishment of decision-making supporting system based on power trading and complying the principle of "**business motivated**". From the view of general structure, the core business decision-making supporting system— **technical supporting system of energy-efficient power generation scheduling**, could be gradually established and improved.

The framework of energy-efficient power generation scheduling constructed the business model of energy-efficient power generation scheduling based on the experiences from pilot provinces. It is part of modernized electric power transaction system.

The framework of technical supporting system of energy-efficient scheduling is shown in Figure 3-2.

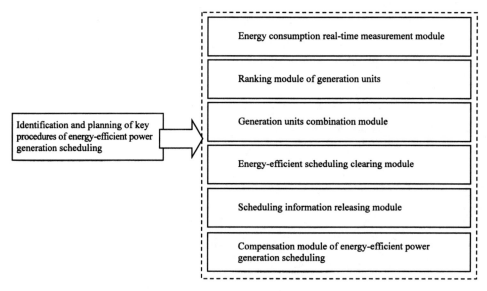

Figure 3-2   Framework of energy-efficient scheduling

3.2   Constitution of energy-efficient power generation information system

3.2.1   Energy efficiency declaration system

This system mainly provides basis reference for local government to prepare sorting table of generating units. It is based on coal quality and units load data from power plants.

Table 3-1   Sorting table of energy-efficient power generation scheduling
of provincial power grid

Unit type	Name	Capacity	Service year	Ranked position
Renewable energy based units				
Hydropower units				
Nuclear units				
Gas-fired units				
Coal-fired units				

(1) Information platform of EEPGS

This is online communication platform for energy-efficient power generation scheduling. Main functions include communication among power plants and power grids, data declaration, data inquiring and service window. The method used to develop this system is Object Oriented Method. The object includes data and processing of data. The core technology is to describe relations and interactions among all objects involved into the energy-efficient power generation scheduling, to achieve simulation and analysis of reality. The major content includes: system segmentation, object recognition, abstract definition of the object, object-oriented modeling, object module and interface technology, integrated design of overall system. The design is in line with the principle of abstraction, information hiding and modular.

(2) software application framework of technical supporting system of energy-efficient power generations scheduling

This is a function integration of energy-efficient power generation scheduling software.

Information platform of energy-efficient power generation scheduling will integrate Energy Management System (EMS), Distributed power control, Wide Area Measurement System (WAMS), grid and micro-grid and Dynamic Stability Analysis (DSA). Through development of Geography Information System (GIS), station monitoring visible and remote viewing technologies, integration, monitoring, emergency response and graphic display can be achieved to improve energy-efficient power generation scheduling from the market-oriented, digital, intelligent point of view.

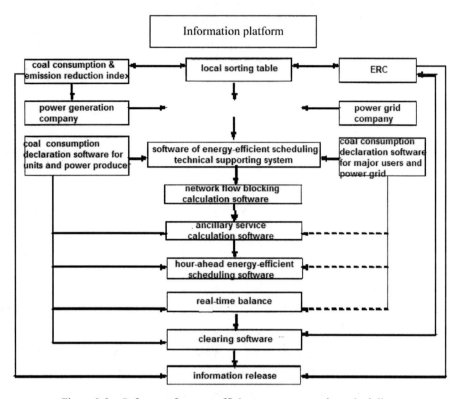

Figure 3-3   Software of energy-efficient power generation scheduling

All software in above figure is necessary and forms the software platform to complete energy-efficient power generation scheduling. Their functions and technical requirements include:

> openness;

> maintainable, in line with changes of primary rules of energy-efficient power generation scheduling and market trading rules;

> scalable, to adapt development of grid dispatching technology and development of energy conservation rules;

> integrity and consistency, to ensure the integrity of data and transactions, and consistency of data;

> high reliability of system;

> high security of system;

> high efficiency of system;

> client zero maintenance.

(3) Business content of energy-efficient power generation scheduling and its information

framework

Compared with conventional scheduling rule, the energy-efficient scheduling is different in terms of procedure, sorting, benefits of both transaction sides and units combination planning.

Main hardware and software of energy-efficient power generation scheduling are shown in Figure 3-4.

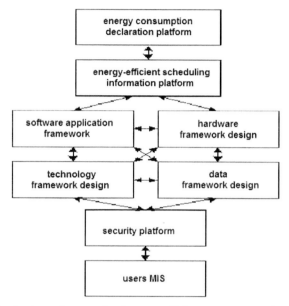

Figure 3-4   Business framework of energy-efficient power generation scheduling

(4) Information framework of energy-efficient power generation scheduling

Energy-efficient power generation scheduling is re-construction of dispatching business in power grid. The technical framework of energy-efficient scheduling is generated under a new scheduling mode. Based on the conventional scheduling mode, plus new sorting rule based on energy-efficient scheduling, the dispatching information of power grid is no longer even-load allocation, but integrated the price factor—real-time price in fuel market. It requires supports from grid dispatching technical system.

As information collection and real-time online dispatching management system oriented primary business of power transaction, energy-efficient power generation scheduling system emphasizes on the integration of power automation technology, Web network technology, SOA technology and Business Process Reset (BPR) mode, hence resulting in the data processing capability of millions of data frames, even thousands of millions data frames by each system per day through multiple communication channels with the compatible and parallel cooperation technologies. Power intelligent analysis of transaction information can be achieved through power purchase by major users and real-time operation status of power generation units.

(5) Technology framework of EEPGS

The primary platform to manage EEPGS is technical supporting system, to carry out large-scale analysis, calculation and management to achieve coordinated unified EEPGS work. The technical framework design uses advanced J2EE technology and multi-level Browser / Server(B / S) structure. Through dynamic software and component technology, multi-layer technical system design is achieved by using shared data mode, in accordance with interface control layer, business logic layer and data layer. Database servers and application servers are constructed through integrated enterprise-level platform and network-based computer platform. Users can realize remote login through browsers, to achieve coordinated work of EEPGS business components at decision-making supporting system. Therefore, various power sources (coal-fired power, hydropower, wind power, solar power and others) and multi-level users (UHV, high-voltage, medium-voltage and low-voltage) can collect real-time production, transaction and investment information through this technical system, to achieve data sharing and information standardization, automation, scientific-grounded and agility, to meet multi-participants ' requirements, to adapt modern, large-scale, real-time and efficient power supply.

System function includes:

➢ **System management.** Focusing on user security management, data authorization management, function authorization management, power grid management and system parameters management. Relevant data and function operation scale should be assigned to different users for safety login and use of system.

➢ **Primary information management**. The data of primary information management includes three aspects:

① Inquire and display the hierarchy of energy-efficient power generation scheduling transaction model and relevant information of the grid scheduling model.

② Query and maintenance of sorting data of power plants.

③ Provide integrated management platform to achieve query and maintenance function for energy-saving and particular data. Main function includes data query, editing and graphic production and output.

➢ **Power generation sorting and transaction data analysis.** This function mode can provide comparison analysis of sorting index of energy-efficient power generation scheduling, analysis of grid stability and correlation analysis of transmission congestion. It can also efficiently and easily find out the historical transaction rule to improve transaction quality and accuracy and provide technical supports for power transaction and clearing.

- ➤ **Market-matching transaction management of EEPGS.** The market-matching transaction is the core module of EEPGS dispatch system. It includes EEPGS long-term power generation planning, day-ahead EEPGS transaction and market-matching in real-time EEPGS market. Data processing includes: coal consumption sorting, transaction submitted price, deal price and so on. Besides, it also provides dispatch procedure related time setting, planning information setting, coal consumption correction, inquiry of sorting results, auto-sorting strategy setting, adaptive training strategy setting, EEPGS blog and other functions.

- ➤ **Post evaluation of EEPGS.** This function mode applies multi-dimensional visualization technology to carry out post evaluation of data quality, load stability, prediction method application, prediction accuracy indictors and historical performance of recommended program to find out the weak link in forecasting and optimize program.

- ➤ **Reporting and assessment management.** Firstly, to achieve duty allocation, transaction results coordination, regulation setting, procedure monitoring, status supervision, free sorting and free assessment setting, assessment index setting and other functions based on management procedure; secondly, to provide accuracy, comparison and index ranking of daily, weekly, monthly, quarterly and annual power generation price, transaction quotas of power producers who participate EEPGS; thirdly, to achieve EEPGS active and reactive predicted results release.

3.2.2 Decision-making supporting system of energy-efficient power generation scheduling

Scheduling management of power system has related hierarchical mechanisms, including: interconnected grid layer, regional power grid layer and provincial power grid layer. Each layer provides services for above layer so that it can play the function of decision-making and show its structural characteristics.

The energy-efficient scheduling is different from conventional scheduling in information transformation. As units are sorted by energy efficiency, the order of on-grid units is changeable, which is similar with market auction. Therefore, to implement energy-efficient scheduling, only management information system is not enough, the support from intelligent expert system is needed as well.

The main task of power grid scheduling is dispatching on-grid power generation quotas of generating units according to power generation planning. The management information system of regular scheduling is show in Figure 3-5.

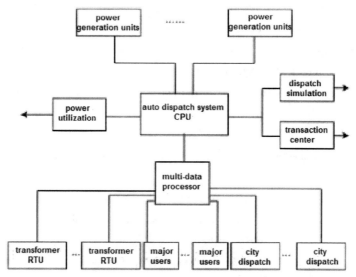

Figure 3-5   Automation system structure of regional power grid scheduling

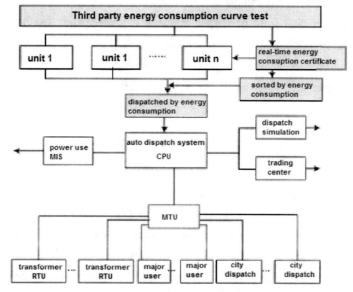

Figure 3-6   Automation system structure of regional energy-efficient scheduling based on coal-consumption sorting

Management information system of energy-efficient power generation scheduling is a system structure of new grid scheduling and decision-making supporting of power generation transaction. It changed a lot based on original management information system of power grid scheduling. Main adjustment content is adding an interface of sorting by coal consumption—"energy conservation sorting mechanism", for on-grid generating units.

After the introduction of energy-efficient power generation scheduling rule, the energy conservation transaction can only be achieved under premise of various standard power grid model, computer, network and other new technologies.

The same type of thermal power units are sorted according to their real-time energy consumption value. Therefore, the business procedure of energy-efficient scheduling will be sequentially conversed based on business content in accordance with the direction of energy-efficient scheduling information flow.

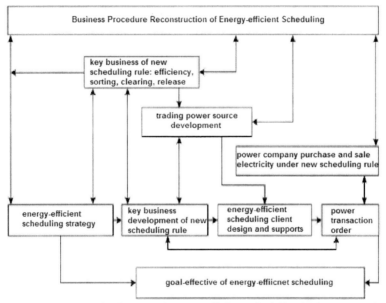

Figure 3-7   Schematic diagram of energy-efficient scheduling business procedure
reconstruction strategy

3.2.3 Decision-making information supporting system of energy-efficient power generation scheduling

Main function of IT platform of energy-efficient scheduling is to assist decision-making information supporting system of energy-efficient scheduling. The IT platform is the "physical platform" to implement energy-efficient scheduling, consisting computer hardware and software. It needs the "chemical platform", such as operation rules of energy-efficient scheduling to instruct, control and coordinate. Besides, a "regulation platform", as commercial operation rules of energy-efficient power generation scheduling, is needed as well, which refers to related implementation regulations of energy-efficient scheduling

(1) Information technology platform of energy-efficient power generation scheduling

The IT platform of energy-efficient scheduling plays the role of hardware to achieve sorting, transaction, pricing and clearing of energy-efficient scheduling. It is power grid scheduling task based on the current power generation planning implemented through related regulations issued by the government.

The basic function of the IT platform is to prepare generating units sorting table based on energy consumption index. Therefore, the IT platform based on coal consumption value is the physical premise of implementation of energy-efficient power generation scheduling.

——Hardware platform of energy-efficient power generation scheduling

Critical servers or processing host of energy-efficient scheduling requires high reliability of system, technologies are included as following:

➢ CPU

①Disk array, cluster computing and fault-tolerance technology. This technology provides safeguarding for key business of large-scale and complicated energy-efficient scheduling. A well-designed cluster computing and fault-tolerance system has continuity of client applications. The technology has no special requirements to hardware settings of backup servers. It also supports share of disk arrays and expansion of pure software's fault-tolerance function with reliable performance.

② Fiber Channel Technology (FCT) and Storage Area Network (SAN). By using FCT technology, RAID can be connected to back-end server, to provide high broadband and transmission rate.

➢ Operation system of energy-efficient power generation scheduling

The openness of the third-generation scheduling automation system relies on the openness of UNIX operation system. The operation system required for energy-efficient power generation scheduling is based on key business of energy-efficient scheduling.

——Software platform of energy-efficient power generation scheduling

Based on the conventional scheduling software, the CSG achieved technical innovations in development technical supporting system. Related software includes:

Guizhou Power Grid:

➢ National invention patent on coal quality industrial ingredient of power plant and thermal value calibration method

➢ National invention patent on determination algorithm for effective desulfurization of coal-fired generating unit of thermal power plant

➢ National invention patent on computing method for desulfurization on-grid power quantity of thermal plant

➢ National software copyright of timely computing analysis of reservoir regulation and its application system of Guizhou power grid

➢ National software copyright of technical software of thermal power regulation of Guizhou power grid

➢ National software copyright of upload evaluation technical supporting platform of load and equipment maintenance of Guizhou power grid

➢ National software copyright of short-term economic operation scheduling system of Guizhou power grid

➢ National software copyright of TPRI-DCCSS general data collection communication software

➢ National software copyright of coal consumption online monitoring system of energy-efficient power generation scheduling

Guangdong Power Grid:

➢ National software copyright of energy-efficient power generation technology and side technology supporting and information management system

➢ National invention patent on method of connection between designed coal consumption value and pollutant emission level and equivalent load of actual power generation during units load allocation

And other general and specific software.

(2) Regulation platform of energy-efficient power generation scheduling[4]

According to the content of construction report of energy-efficient power generation scheduling, regulation platform of energy-efficient scheduling includes:

1) Regulations at national level

➢ Energy-efficient power generation scheduling approach (trial)

➢ Working program of energy-efficient power generation scheduling pilots

➢ Guideline for implementation of energy-efficient power generation scheduling (trial)

➢ Management approach of desulfurization power price and desulfurization equipment operation of coal-fired units (trial)

➢ Information dissemination approach of energy-efficient power generation scheduling (trial)

---

4   According to pilot summary meeting of EEPGS in China Southern Power Grid

> Notice regarding on economic compensation of energy-efficient power generation pilots

2) Regulations at provincial level

> Flue gas desulfurization real-time monitoring and desulfurization power quotas evaluation system operation management regulations for on-grid coal-fired units (trial)

> Operation management approach of coal consumption online monitoring system of energy-efficient power generation scheduling (trial)

> Leading group foundation document of energy-efficient power generation scheduling

> Implementation program of energy-efficient power generation scheduling

> Units sorting table of energy-efficient power generation scheduling

> Economic compensation approach of energy-efficient power generation scheduling

3) Regulations at power grid level

> Guidelines of technical supporting system construction of energy-efficient power generation scheduling for China Southern Power Grid

> Technical regulation on flue gas desulfurization real-time monitoring of on-grid coal-fired units of China Southern Power Grid (trial)

> Technical regulation on coal consumption online monitoring system for energy-efficient power generation scheduling in Guizhou

> Evaluation regulation on energy-efficient power generation scheduling statistic in China Southern Power Grid

(3) Overall framework of decision-making supporting system (IT) of energy-efficient power generation scheduling

The overall framework of IT includes scheduling business procedures, application framework, technical framework, data framework, physical framework and security framework. Business procedure is the basis of IT construction. The function of IT system should be integration of power supply chain in the rules of energy-efficient power generation scheduling.

(4) Management structure of grid hierarchical scheduling

The grid hierarchical scheduling management structure is based on grid connection of "units - power grids", provincial, regional "grid – grid" connection and joint operation of "hydropower – thermal power", "thermal power – wind power - hydropower" and other power sources. The power grid was used to responsible for power transmission and power distribution, and the power scheduling platform was constructed by hierarchical structure at different voltage standard, 1000kV—750kV—500kV—220kV—110kV—35kV—10kV. The energy-efficient power generation scheduling is achieved through interconnected power grid, large-scale power grid, provincial power grid and regional power grid, respectively with UHV and EHV of 1000kV—750kV—500kV by national interconnected power grid dispatch agency, 220kV—110kV—35kV managed by provincial power grid dispatch agency and 10kV by city, regional and county level dispatch agency.

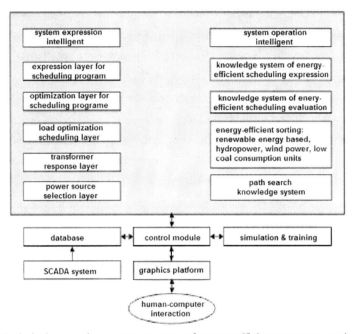

Figure 3-8   Technical supporting system structure of energy-efficient power generation scheduling

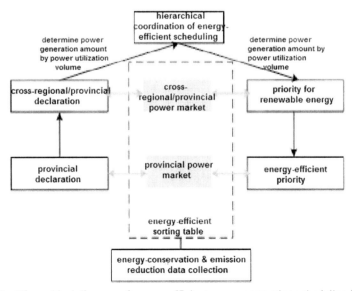

Figure 3-9   Hierarchical diagram of energy-efficient power generation scheduling information

(5) Reporting and operation data framework of energy-efficient power generation scheduling

With the rapid development of power industry in recent years, massive power grid data has been accumulated in power dispatch agencies. Due to those data is generated under conventional scheduling rule and it is specifically used for dispatch agencies, it is isolated and is not in accordance with energy-efficient scheduling rule. Therefore, the data should be classified by dividing into offline and online two categories, then form different databases.

1) Offline data system of energy-efficient power generation scheduling

Offline data of energy-efficient scheduling refers to data which cannot directly determine the operation and result of energy-efficient scheduling, such as meteorological data, personnel data, instant message and etc.

2) Online data system of energy-efficient power generation scheduling

Sorting table based on designed coal consumption of units, coal consumption curve of thermal units, integrated energy-conservation optimized power generation scheduling mode. The regulation of those data should be in line with following technical guidelines.

➤   Inquiry database of technical regulations

GB8117. 1 – 1987 "Regulation on thermal performance acceptance test of steam turbine"

ASME PTC6 – 1996 "Regulation on performance test of steam turbine" American Society of Mechanical Engineers; and so on.

➤   Third-party certification database of coal consumption

By using SCADA data collection system and anti-balance method, this type of data can remove impacts from environment and human beings and accurately reflect technical performance of equipments. Due to this type of data was used to evaluate equipment manufacturer and direct energy-conservation improvement for power plants, different objectives would refer to different correction projects. Therefore, when the data is used in energy-efficient power generation scheduling, specific correction index should be removed to reflect the actual online real-time energy consumption level of units.

Coal consumption calculated by anti-balance method is based on total concentration of heat of steam boiler and anti-balance efficiency of boiler, to firstly calculate consumption of coal equivalent and then raw coal quantity:

Coal equivalent consumption of boiler = heat output of boiler/(anti-balance efficiency of boiler × heat content of coal equivalent)

Raw coal quantity consumed by boiler = coal equivalent quantity consumed by boiler / raw coal calorific capacity

where, heat output of boiler = (total weight of boiler superheaded steam – total weight of) × (heat content of superheaded steam of boiler exit – heat content of supplied water

-82-

in the boiler) +total weight of warm water of boiler × (heat content of superheaded steam of boiler outlet - heat content of supplied water in the boiler) + total weight of sewage water × (heat content of sewage water from boiler - heat content of supplied water in the boiler) + self-use steam quantity of boiler × (heat content of self-use steam of boiler – heat content of supplied water) ± others (other intake or outtake heat not included)

The generation of this type of data is coal consumption curve based on the load curve. Due to load rate is different for each unit, steam quantity consumed in each power generation station is different. Therefore, each coal-fired unit should construct its coal consumption curve during energy-efficient power generation scheduling, then power generation quota is allocated to each unit based on the coal consumption curve with principle of equal incremental.

According to function requirement analysis of energy-efficient power generation scheduling, function structure of proposed management information system of energy-efficient scheduling is shown in Figure 3-10.

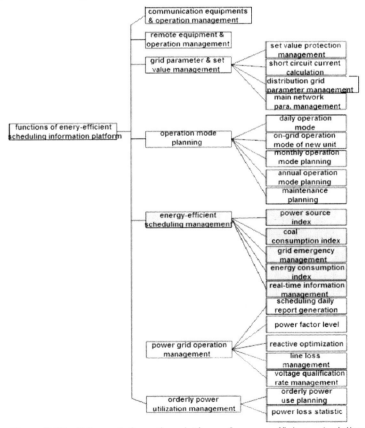

Figure 3-10   Primary information platform of energy-efficient scheduling

3.3 Inputs of energy-efficient power generation scheduling

3.3.1 Personnel safeguard for energy-efficient scheduling technology

The successfully experience in Guizhou pilot demonstrated that the implementation of energy-efficient scheduling is based on establishment of organizational safeguarding, which composed by personnels. Therefore, a team of independent system operator (ISO) should be established to promote and improve energy-efficient scheduling. The energy-efficient scheduling aims to energy conservation and maintain independent of trading. Only in this way, the objective, guideline and principles can be strictly implemented.

Training and re-training should be conducted for staff working in Grid companies. Position in the power dispatching should be clarified for dispatching staff, including technicians at provincial, municipal and inter-provincial levels. Working field should also be clarified, including power system, thermal system, hydraulic water system or trading calculation. Different human resource training should be prepared for staff at different position with specific training approaches. Mid-term and long-term prediction should be prepared for energy-efficient scheduling promotion.

3.3.2 Material safeguard for energy-efficient scheduling technology

Material resource is the basis for promotion of energy-efficient scheduling based on above technical systems and equipments.

Material basis mainly includes: hardware, software, data, user and system, analyzed as below:

(1) Hardware

Hardware resources include all hardware equipments needed for energy-efficient scheduling. In addition to dispatching system equipments such as EMS and AGC, other hardware equipments include facilities of coal-consumption online monitoring system, flue gas desulfurization system and others to ensure the operation, data storage and processing of software. Detailed content of hardware can be found in Table 3-2.

Table 3-2   Main hardware resource of energy-efficient scheduling

Items	Description
Terminals (RTU)	Involved dispatching terminals of power generation, power transformer, regional grid, provincial grid and local grid.
User serves	Serves providing communication management for multi users and provincial grid companies.

Items	Description
Dispatching central serves	Serves managed by grid communication management center, which provide dispatching service for the whole Grid.
Network management equipments	Equipments conducting network management, located in network management center.
Backbone interconnection equipments	Equipments used for backbone interconnection, such as switches.
Internal interconnection equipments	Equipments used to connect provincial interconnection equipments, as well as interconnect equipments between local power grid and among power plants, such as, routers, security access routers, secure routers and network encryption machine.
Communication equipments	Various adaptation equipments, modem, switch, etc.
Connections and cabling system	All connections and cables and etc.
Dispatching management center	Equipments used to manage other equipments during dispatching period, such as password distribution equipment, certificates center and etc.
Trade clearing center	A series of trade clearing system connected to the bank is necessary hardware equipments.
Security equipments	All equipments to ensure grid security, such as cipher machine, security cards and etc.

(2) Software

As for pilot work of China Southern Power Grid, innovation in software integration has been conducted. The Guangdong Grid established a new generation system of network-based energy-efficient power generation technology supporting and information management.

1) Content of software resources

The software system covers core functions such as preparation of energy-efficient power generation scheduling plan, energy efficiency assessment, system load forecasting, bus load forecasting, net loss analysis and revision, security check and etc. The software established mathematical model for energy-efficient scheduling, and proposed efficient solving approaches. The software system provides the method of daily power generation planning, and gradually optimizes it in a feasible range, which resolved the optimization problem at a large scale with complicated constraints.

The project proposed a highly adaptable energy-efficient scheduling hierarchical optimization method. Based on joint optimization of primary and auxiliary targets, multi goals are proposed including NDRC-order, lowest coal consumption, lowest total power purchase cost, balanced dispatching, and maximum energy conservation effect. Through balance of primary and auxiliary objectives, hierarchical optimization planning for energy-efficient scheduling can be achieved, to fully meet the diversity and flexibility of preparation of energy-efficient scheduling plan.

Major technical and economic indicators of this system include:

① The preparation of power generation planning for energy-efficient power generation scheduling should consider power transmission capacity of grid network, peak shaving capacity of power grid, units maintenance planning, power transmission line loss and other constraint conditions. The planning maker could set a variety of boundary conditions, initialization conditions, operation status of units and etc. to prepare non-constraint and constraint power purchase plan and optimize day-ahead power generation plan through parameter control, also provide blocking analysis information.

② Predict monthly and daily load curve and load curve of different time in a week.

③ The most appropriate historical data and forecasting methods can be selected through pattern recognition to predict bus load. Based on the bus load forecasting, deviation correction to forecasting result can be carried out by non-linear analysis model. Considered the large quantity of bus line and large forecasting workload, pattern recognition and follow-up forecasting procedures are automated. Meanwhile, manual intervention interface is designed for the whole procedure.

④ Scientific and systematic evaluation indicator system of energy-efficient power generation scheduling can be used to analyze benefits of coal saving and energy conservation by setting the load rate of units as key property; to analyze impacts on cost and coal consumption rate by different type of unit and find out the key impact factors; to judge the dispatching fairness of power generation planning according to scheduling optimality of units based on coal consumption level.

⑤ When the WEB server is used by 30 users at the same time (normal condition), any user using the same LAN as WEB server browers any page of the server, the average response time of the whole system should be less than 10 seconds to ensure high availability of the system.

⑥ The data scale which can be processed in the system is: maximum generating units: 300; lines: 500; nodes: 500.

⑦ Annual utilization rate of the system $\geq 99\%$.

2) Material resources

Independent software products which can be purchased separately (or services and supports could be independently obtained) will be used to process all data related to energy-efficient scheduling. Major software resource can be found in Table 3-3.

Table 3-3 Main software resources

Items	Description	Related hardware resources
Windows	Windows system for every user's PC	User's PC, network management equipments
Windows NT	Windows NT system for some user's PC and servers	Servers of network center and user's servers and network management equipments
Internet public service software	Mainly including DNS, WWW, FTP, E-mail List, News and etc.	Servers of network center and user's servers
Database management system	Specific and general database system	Servers of network center and user's servers
Personal software	Such as word processing software, internet explorer, E-mail software and etc.	User's PC
Commissioned developed application system	Contractor developed application system, such as coal consumption online monitoring system and etc.	Servers of network center and user's servers, user's PC
Independent developed application system	Independent developed application system, such as, energy conservation benefits evaluation subsystem and information dissemination subsystem.	Servers of network center and user's servers, user's PC
Specific software for dispatching	Specific software, such as sorting subsystem and dispatching combination subsystem, security and network loss dispatching system	Servers of network center and user's servers, user's PC
Grid management system	network management software used by network energy-efficient scheduling	Network management equipments
Specific embedded software	Software used in all systems	External networking equipment, internal networking equipments, communication equipments and security equipments.
NOTES	Software platform such as OA	Servers of network center and user's servers, user's PC

(3) Data resources

The operation and management of power grid in the PRC is characterized as typical hierarchical partitioning. The implementation department of energy-efficient power generation scheduling is dispatching center of power grid, which is divided into several departments. With the implementation of energy-efficient power generation scheduling, related power generation data needs to be integrated, as well as grid transmission information. For the management of all data, systems integration should be conducted to operation decision-making of energy-efficient scheduling system.

Data resource of energy-efficient power generation scheduling refers to data which is automatically or manually generated during the operation period of energy-efficient power generation scheduling. This resource is accumulated over time. The value of data resource is to promote energy-efficient power generation scheduling. The ultimate goal of energy-efficient power generation scheduling is to ensure safety and smooth implementation of new scheduling rule and achieve energy conservation.

To facilitate decision making analysis of energy-efficient scheduling, in addition to identification of data resources, their contribution to energy conservation should also be identified; due to higher contribution data has greater value.

Contribution to energy conservation can be divided into following grades:

1) **Direct contribution to energy conservation.** Including coal consumption data, flue gas desulfurization data; constraint data of energy conservation of national, provincial and regional Grid.

2) **Moderate contribution to energy conservation.** Working data of internal power grid. Such as, load forecast data, power flow data and coal consumption online monitoring data.

3) **Indirect contribution data to energy conservation.** Such as network loss data, manual data, equipment maintenance cost data.

Meanwhile, the data resource of energy-efficient power generation scheduling includes public data, data for inner users, confidential data and highly confidential data. Public data refers to energy conservation and pollutant emission data which can be public or used in dispatching activities, for example, energy conservation strategy of power plants, periodic energy conservation performance, leaders' speech, flue gas desulfurization monitoring data of local industries; data for inner users refers to periodic investment and planning of energy-efficient power generation scheduling of local industries/companies, periodic cost and performance of energy conservation, coal consumption monitoring data of each period; confidential data refers to business management, financial data and salary data related to energy-efficient power generation scheduling.

Data resources related to energy-efficient power generation scheduling can be found in Table 3-4.

Table 3-4   Main data resources

Resource type	Description	Characteristic
Energy-efficient scheduling publicity data	general publicity material, public dispatching performance data	Public
Policies, regulations and other documents data	national energy conservation policies, laws and regulations; Documents of energy-efficient scheduling at grid level.	Public
Released data of energy-efficient scheduling	dispatching operation data	internal or half-public to participants
Grid operation data	Specific and general grid operation data	Internal
References	Published paper and book related to energy-efficient scheduling	Public
	Internal documents, such as: working summary, documents from on-site meetings	Internal
	experience summary and performance assessment	Internal

Resource type	Description	Characteristic
Scientific and technological information	International or domestic specific technology, such as patent	Internal, confidential
Market information	application system developed by contractors, such as coal consumption online monitoring system and etc.	Confidential
Operation data of energy-efficient scheduling system	data processed by independent developed application, such as, energy conservation benefits evaluation data, information dissemination data	Internal

(4) Users

The user resource of energy-efficient scheduling is the premise of high efficient and safety implementation of working objectives. Users' participation is the key point of dispatching promotion. The main implementation measures of energy-efficient scheduling aims to upstream users—power generation companies. Their participation in energy-efficient scheduling mainly includes:

1) Give dispatching priority to renewable resources in accordance with "Energy-efficient power generation scheduling approach";

2) Prepare ranking list in accordance with energy consumption of generator and pollution level of emissions;

3) Dispatch fossil energy resource as power generation resource in turn;

4) Mandatorily decommission inefficient small generators according to administrative measures;

5) Conduct power generation rights trading, so that efficient generators could gain more power generation quota.

All users involved into energy-efficient scheduling have their own specific working scope. They all have rights to operate and maintain network system within their working scope. Also professional training should be arranged in energy-efficient scheduling field.

(5) Supporting systems

To ensure normal and safety operation of energy-efficient scheduling, a number of safeguarding systems are needed, mainly including:

1) Communication system

2) Weather forecasting system. Consider the load change trends.

3) Demand side management system.

4) Policy consulting.

5) Performance assessment system

Overall, material resource is the basis of energy-efficient scheduling. By development technical supporting platform, material resource can be used to in actual production period, to improve the scientific and decision-making level of the power grid.

3.3.3 Financial safeguard for energy-efficient scheduling technology

Establishment of financial supporting system for energy-efficient scheduling is necessary for dispatching promotion.

According to the investment scale evaluation for pilot dispatching in Guizhou province, the total investment of Guizhou Grid is around CNY10 million, of which hardware accounts for a big part. Total investment of China Southern Power Grid reached to CNY 50~80 million. The total national investment reached to around CNY300~500 million.

As for the benefit analysis of energy-efficient power generation scheduling, in 2008, in total 426,000 tons coal equivalent was saved, economic benefit reached to CNY 260 million, resulting in reduction of CO_2 emission of 1.333 million tons and 778,000 tons of SO_2[5]. Therefore, Energy-efficient scheduling could bring benefits to power generation companies in saving cost of fuel, as well as environmental benefits. However, as the implementation body for energy-efficient scheduling, the benefit for grid network itself is not significant. Especially when renewable energy connects to the power grid, the investment to the grid might be increased due to increased investment of information management systems. Therefore, financial safeguarding planning should be prepared, otherwise, the expectation results would be difficult to achieve.

With consideration of incentive, a series supporting system should be established to promote energy-efficient power generation scheduling in terms of financial investment planning, system construction, operation, organization institution, legal system and public advertisement.

Therefore, detailed and systematic planning for promotion of energy-efficient scheduling around the Grid or at national level should be prepared according to different power load period, through safeguarding in design of standards and procedure, real-time monitoring, equipment and environment monitoring and online supporting.

3.3.4 Technical training of energy-efficient scheduling

Preparation of technical training program is the basis of promotion and implementation of energy-efficient scheduling.

---

5   Data source: documents compiled in Guizhou in 2010

(1) Training content

1) Training content framework

Training in policy aims to raise awareness of decision-maker, manager and technician in energy efficiency, energy conservation, emission reduction, fair trading, generator sorting, coal consumption and pricing, to provide them policy-based guidance during practical period.

Training content in this part includes: energy efficiency module, policies and regulations in energy conservation, principles of generators synchronization, trade contract of power market, principles of information dissemination and etc,.

Power plant operation management includes: international and domestic power trading, operation and management of power plants and grid companies.

Technology of energy-efficient scheduling includes: sorting method analysis, coal consumption of coal-fired generator and its sorting method, power generation plan, trading clearing, technical supporting system and its operation, summarized as Table 3-5.

Table 3-5　Training content of energy-efficient scheduling

Training content framework	Policy	Related policies and regulations	
		Energy conservation law	
		"openness, fairness and impartiality" dispatching regulatory	
		Energy efficiency training	
	Management training	Power system reform	
		Power plant operation management	
		Energy efficiency management	
		Power price and cost	
		Environmental engineering of desulfurization and denitrification	
	Technical training	Grid integration rules	
		Power trading mode	
		Dispatching plan	Annual plan
			Monthly plan
			Day-ahead plan
			Balance plan
		Coal consumption management information system	
		Renewable energy power generation technology	Large scale wind power generation technology
			Large-scale solar power generation technology
		Professional knowledge training	Power system and its automation
			Hydropower automation
			Systematic project
			Computer network project
			Management information system
		Transaction clearing	

Of which, the energy-efficient power generation scheduling policy module is compiled all policies related to energy-efficient scheduling. The policies can be divided into: documents issued by the State Council, documents issued by the NDRC, documents issued by State Electricity Regulatory Commission, documents from regional power grid, and related documents issued by provincial and provincial power grid.

The power system reform module should introduce the basic content of international power market and international energy-conservation work, and display the background and significance of power conservation in the PRC, its method and measures, theory and practice, difficulty and barrier, its future and prospects, to train qualified personnel for the implementation of energy-efficient power generation scheduling.

(2) Power system and its automation

Training of power system and its automation needs involved trainers have related knowledge in power grid operation mode, basic principle, capacity and load adjustment method, power market, trading principles, basic analysis and calculation method of power grid.

For the theoretical part, they should have a basic knowledge of dispatching information, dispatching hierarchical control, remote technology of power system, automation system, dispatching automation function, automation of power distribution and so on.

Power system communication, radio communication, carrier communication, fiber optic communication, data communication, intelligent grid technology and etc,.

In the practical section, the technical guidance of energy-efficient scheduling should also be included into training content, including:

a. applicable scope of energy-efficient scheduling.

b. sorting of generator combination.

c. generator combination program.

d. economic dispatch and security constraints.

e. equipment maintenance, peak shaving, frequency regulation and reserve capacity arrangement (ancillary services).

f. information disclosure and supervision.

(3) Water resources and hydropower automation

Hydropower dispatching should be given priority in sorting and dispatching. Training content related to hydropower should include basic knowledge of hydropower automation, benefits to energy conservation, integrated utilization of resources, peak shaving, safety and economically operation.

(4) Thermal power project

The Power Grid in China, eventhough the sorting table is prepared, and the priority is given to hydropower and other renewable energy. Thermal power generation still accounts for a big proportion. The dispatching sorting need be prepared according to coal consumption. Therefore, technical capacity of dispatching should include capacity building of thermal power project.

Contents include:

1) Coal consumption. Construction and operation of coal consumption online monitoring system;

2) Coal-fire generator dispatching. Load rate of coal-fire generator, load curve design of power generation, peak shaving of coal-fire generator.

(5) Systematic project

Power technology is no longer only an engineering technology, it became a complicated integration body with power, information, human resource, market, climate, and even politics and diplomatic field. How to design and operate this system while avoid interference to other social systems. System engineering methodology should be the best choice, to reasonable operate this complicated power system.

The core content of energy-efficient scheduling is energy conservation and emission reduction. Besides, it also includes decision-making factors of hydropower system, coal-fired power generation system and new energy system. See Table 3-6.

Table 3-6   Data structure of dispatching in hydropower system

Management operation system of daily dispatching plan of hydropower plants	Annual power generation plan	Hydrological data input
		Indicator of hydropower annual generation plan
		Hydropower plant annual production plan
		Hydropower plant maintenance plan
		Sorting index
	Integrated information dissemination database	Reservoir operation data
		hydropower plants generated power statistics
		Phasing power statistics
		Power balance and power utilization rate by power plants
		Water consumption rate
		Equipment utilization hours
		Output reports
	Query	Hydrological data
		Power generation plan
		Trading information dissemination
	System maintenance	Data backup
		Data recovery

Along with the implementation of energy-efficient scheduling, renewable resources get involved, such as hydropower. Coal-fired generators still need to be integrated for power generation. Related power generation plan should be established aiming to thermal power generator sorting.

Table 3-7 Data structure of dispatching in thermal power system

Management operation system of daily dispatching plan of thermal power plants	Annual integrated power generation plan	Coal consumption data input
		Generating unit power generation plan
		Turbine operation parameters
		Boiler operation parameters
		Indicators of annual power generation plan of thermal power plants
		Forecasting indicator of annual production plan
		Planned maintenance capacity
		Sorting index of coal consumption
	Integrated information dissemination database	Thermal power plants generated power data
		Generation-supply power balance
		Coal consumption rate and coal consumption statistics
		Heat supply statistics
		Units operation statistics
		Sorting dispatching rate
		Output reports
	Query	Hydrological data
		Power generation plan
		Trading information dissemination
	System maintenance	Data backup
		Data recovery

(6) Computer and management information engineering

With new dispatching rules, informatization of power industry would be greatly promoted.

Computer technology, internet, database, and related hardware are necessary for the construction of energy-efficient scheduling system.

Based on the pilot experience, during the promotion of energy-efficient power generation scheduling, one important work is whether the function of technical supporting system is well known by dispatching technicians or no. Three major problems are summarized as below:

Firstly, lack of technicians who are familiar with dispatching;

Secondly, different regions have their own energy source structure, which directly impact the sorting method, function of technical supporting system;

Thirdly, different participation method has different compensation approach.

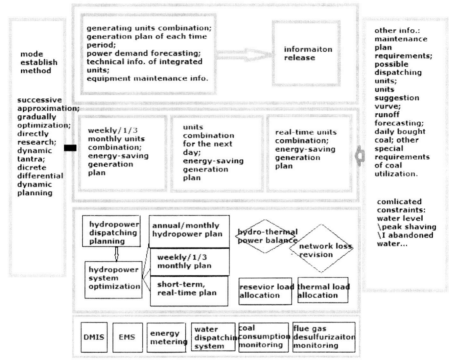

Figure 3-11   Technical system of energy-efficient scheduling

4. Power Market Reform and Energy-Efficient Power Generation Scheduling

4.1 Major measures for power regime reform in the PRC

4.1.1 Break monopoly of investment; separate the government from corporate

At the beginning of reform and opening to the outside world, the power industry started to adjust industrial management regime. Also, in the investment field, the central government tried to cooperate with local government for construction of power plants. Construction of rural small hydropower station was encouraged. A series of reforms have been explored by introduction of international advanced experiences in power industry.

To speed up power construction and actively improve power industry, in 1985, the State Council together with some other departments issued "Interim Provisions regarding on Encouraging Capital-raising for Power Generation and Implementing Multiple Power Pricing Systems", officially, the capital-raising for power generation policies issued; in 1987, policies such as "separation of government from enterprises, provinces as entity, joint power grids, unified scheduling, capital-raising for power generation" and "consideration of local and grid situation" were proposed.

Meanwhile, we should firmly implement reform and opening to the outside world, apply loans from international financial organization, such as World Bank, ADB, Japan Bank of International Corporation and etc, encourage foreign capital investment, multiple cooperation, BOT and other approaches, support listing aboard, aboard contract and technical service, to introduce foreign capital as well as advanced technology, personnel and management experience.

We should promote rapid development of power industry emphasizing on both domestic activities and reform to outside world. From 1985 to 1995, the installed capacity has increased by 1.5 times, which changed the long-term power shortage situation. The PRC successfully developed a way of multiple power generation, multi-channel fund raising with Chinese characteristics.

With multi-channel fund raising, a situation of "power grid is managed by the state, power plants are constructed by multiple investors" has been formed, which proposed new requirement of power system reform. In 1993, power industry department proposed to reform power system in line with principle of "enterprise system reorganization, commercial operation, legally management", thereby Huaneng Group was established, power enterprises commission was founded.

Separation of government from enterprises is a breakthrough in power market reform. By the end of 1996, the State Council established State Power Grid Company, and transfer related duties to National Economic and Trade Commission. From 1998 to 2000, provincial power system was reformed according to related documents, the "separation of governmental from enterprises" was implemented nationalwide. The Power Law was effective on April 1, 1996, which embodies legal system for power industry was formed.

To introduce market mechanism to power industry, the State Power Grid company was founded, to establish modernized enterprises system, and introduce market competition mechanism into power generation. In 1998, related documents issued by the State Council and National Economic and Trade Commission were published, and six provinces/cities carried out reform of "separation of power plant from power grid, on-grid auction", market-oriented power price was proposed, market based theory such as price-direct investment and consuming was proposed. The awareness of "separation of power plant from power grid" reform was raised.

Since 1998, CNY 300 billion was invested to reconstruct rural power grid, reform rural power management system and achieve urban-rural same grid same price. Through this project, the backwards rural power infrastructures were improved, smoothened the relations between county-level power supply company and provincial power company as well as town/township level power station, realized urban-rural same grid same price in most of China.

4.1.2 Separation of power plant from power grid

In order to further establishment of market economic mechanism and reform of monopoly industry, in March 2002, the State Council issued "Power system reform program" (No.[2002]5) based on domestic situation and international experiences. It clearly proposed reform targets "to break monopoly, introduce competition, increase efficiency, reduce cost, improve power price mechanism, optimize resource allocation, promote power development, promote nationlwide power connection, separate government from enterprises, construct a fair, orderly competition to promote healthy development of power market system".

According to requirements on power market construction, the reform emphasis in 2002 includes:

——Separation of power plant from power grid. Distinguished the monopoly and competition, the power generation capital of State Power Grid was separated from power grid capital, and five power generation groups were established. Required companies which provide ancillary services, such as power design, maintenance and construction, should be separated from power grid companies. A fairness competition situation was formed.

——Construction of regional power grid company. To further optimize resources, State Power Grid and China Southern Power grid was established. And five regional power grid companies were established under the State Power Grid, to eliminate independent provincial power grid.

——Grid integration auction. Within operation scale of regional power grid, a or several power dispatching transaction center(s) will be established according to regional power grid structure, load distribution and power tariff level, managed by regional power grid company. Power dispatching transaction center will be opened to the market.

——Power tariff reform. The power tariff was divided into on-grid tariff, power transmission tariff, power distribution tariff and end-users tariff. On-grid tariff includes capacity tariff and energy tariff; power transmission and distribution tariffs are determined by government; end-users tariff should be connected to on-grid tariff.

——Established State Electricity Regulatory Commission. It is directly-controlled by the State Council. Representative authorities are assigned to power dispatching transaction center of regional power grid companies, responsible for power regulatory to protect legal interests of power investors, operators and consumers; establish regulated power operation order and safeguard fair competition in power market.

Power market reform promoted rapid development of power industry; it took only three years to solve power shortage in the PRC. In 2007, installed capacity and power grid construction scale doubled than that of 2002. Competition and service warns of enterprises was raised, and operation efficiency was enhanced annually. Development of energy-efficient renewable energy was greatly encouraged.

4.1.3 Regional power market pilots

Power market construction is the major content of power system reform. Since 2003, power market construction pilot work was carried out in northeast, north and south China.

——Characteristics of power market in northeast China, implemented annual and monthly price auction based on capacity tariff and energy tariff.

——Characteristics of power market in north China, learned experiences from Australia and USA, explored the feasibility of differential price contract and 节点电价.

——Characteristics of power market in south China, combined with west-to-east power transmission, price auction was 模拟 operated in some power plants for some power generation amount.

——Researches on power market construction were also carried out in central, northwest and north China, and proposed some innovation thoughts. For example, power market in central China emphasized on combination of hydropower and thermal power; power market in northwest China focused on direct power purchase by major users and inter-provincial and inter-regional power balance; north China paid more attention on bilateral transaction program design.

Through several years of exploring and practice, regulations, such as "Guidelines of regional power market construction", "Primary rules on power market operation", "Regulations on technical supporting system function of power market", "Power market regulatory approach" and etc,. Related power market construction programs were issued aiming to different power markets. Attempts, such as two-settlement power price, differential power price and auction for some electric energy, were made to enhance market-oriented awareness of power companies.

However, pilot projects of power market reform did not achieve expected results, reasons can be summarized as:

——the market construction reform was chosen under the condition of power shortage; power system faced serious challenge;

——reform supporting measures were not matched, without balanced account and electricity tariff linkage mechanism, resulting in price volatility;

——with constraint of planning economy, if there is no large-scale interest adjustment to change traditional even-load power generation quotas and change power price regulatory, reform barriers would exist.

Overall, under single-purchase condition, due to lack of price-direct mechanism between power plants and users, the market formed back then is not the real power market. Especially, under serious scarcity of power supply, the power tariff after auction once is higher than on-grid power tariff, a thought "reform means price increase" would happen to most users. In this way, it is hard to promote power market reform.

4.1.4 Direct power purchase by major users and bilateral transaction

Combined with international and domestic reform practices, the possibility of large account user direct purchase was discussed. In March 2004, "Interim approach of power users directly purchase from power plants" was issued. It is based on actual situation of power industry in the PRC, learnt from international experiences, in line with power industry development rules to ensure safety operation of power grid and promote reasonable power price mechanism and eventually promote power market construction.

In March 2005, Jilin province officially signed "Power direct purchase contract", achieved the first "point-to-point" direct transaction activity. In November 2006, one power plant in Guangdong province signed direct purchase contracts with six major users, achieved the first "one-to-multiple" direct transaction activity. Based on two direct purchase pilots, in 2009, bilateral or 多边 transaction pilot work carried out in Fujiang and Inner Mongolia.

The significant change was expressed in following aspects:

——extended transaction participants, including qualified power generation companies and industrial users;

——diversity of transaction items, including direct power purchase by major users, power generation right transaction and inter-provincial/regional power transaction;

——transaction method is more flexible, including annual, monthly bilateral transaction, as well as day-ahead market, matchmaking transaction and ancillary service transaction;

——More standardized market transaction, including completed power market construction program, distinguished responsibilities and obligations of power generation company, power grid company and users and clear task divisions for government departments and regulatory entities.

Meanwhile, the power market platform has been continuously strengthened to regulate inter-provincial/regional power transaction. Since 2009, the platform of east China power market achieved direct participation of transaction by power generation enterprises, the market price signals are formed, and power generation resources are further optimized.

4.1.5 Problems need to be further resolved

Entered to the new century, driven by accelerated urbanization and industrialization, the human-resource-environment pressure has been increasing. With the fast growth of power industry, some problems which are against the scientific development theory and harmonization social construction are exposure, such as irrational power source structure, transmission of coal-fired power, pressure of natural resources and environmental protection, backwards construction of power grid, uneven development among regions, uneven rural-urban development, uncompetitive in self-innovation and international market and so on. In general, the high-investment, high-emission and low-efficiency development in power industry limited the primary function of market in resource allocation. As for the fundamental reason, it is the problem of power development regime and mechanism.

The first problem is irrational power market structure. From the international experience, eventhough different country chooses different market pattern, different reform method, but one thing in common, which is market structure with multi-level power sellers and power buyers, and separation of power transmission from power selling and purchasing. However, in the PRC, the separation of power plants from power grid is basically established, but the market structure is still irrational. Related information is not shared between power plant and power grid. Power plant and power users have no selection rights. Due to power suppliers and power consumers cannot directly contact, the real power market would not be formed, neither well organized power market order.

The second problem is that the function of price system is not obvious. For many years, the electricity tariff in the PRC only includes government-approved on-grid power price and sale price. The general progress of power price reform is relatively slow eventhough the reform direction has been clarified. The current power price policy cannot reflect the supply-demand relation, or rarity of resource and environmental cost. Due to blur boundary between power transmission and power distribution, the problem of intercross compensation is serious. So far, an independent power transmission and power distribution price is not established yet. The key factor to implement energy-efficient power generation scheduling, limit blindness development of high-energy consumption industries and resolve the conflict of coal-fired power price is breakthrough of price reform.

The third problem is that independent dispatch and transaction entities. Power grid company is integrated with power transmission, power distribution and power sale business, as well as dispatch and transaction functions. In the power market, the power grid company has to participate the power transaction in one hand; in the other hand, it has to implement the scheduling as well. At this point of view, it is hard to ensure an openness and fairness transaction in the power market. Besides, under this regime, it is also hard for renewable energy to connect the grid and healthy development of distributed power generation.

The fourth problem is backwards shift of government function. To establish new power market operation regime, not only a rational market structure, completed power transaction pattern, sound legal system is needed, but also governmental management regime and effective market regulatory policy. In the PRC, the conventional planning economy has impact the power management regime deeply, the government is get used to establish power price and allocate power generation quotas. The responsibility between government and regulatory agency overlapped, which negatively impacted the power market reform.

4.2 Experience and enlightenment from power market reform in foreign countries

4.2.1 International experience

It has been 30 years since reform happened to power market in the world. Even though the background, method and approach of reform are different, but the reform direction remains the same. There are still some successful experience can be learnt.

(1) Break monopoly status; Separate the competitive business and monopolistic business, to shape a qualified market.

Qualified market structure and market player is the basis of the formation of power market operation mechanism. Generally, power industry is considered as typical monopoly industry, and its monopoly position is regarded as natural monopoly, which is limited to power transmission and distribution grid. Power generation, electricity sale, as well as power design, construction and equipments manufacturing is competitive, they can be separated from monopoly business, to achieve separation of power plant from power grid, transmission from distribution, main service from ancillary service. From the experience of foreign power industry, few auxiliary assets existed, they emphasize on separation of power generation plants and power grid, as well as transmission and distribution. The key work is to separate and restructure power generation and power transmission. For example, the UK, Australia, Brazil and Argentina restructured and reorganized their power transmission and distribution linkage aiming to traditional single-type operational power plants; USA promoted separation of power generation from power transmission; cases in Germany is special, the power transmission is independent; France is typical in single-type operation, requested by EU, it is legally to separate power transmission from power distribution. No matter what approached adopted by them, one common point is that franchise operation should be carried out for power transmission and distribution. Power Transmission Company only provides transmission service without participation of any other power trading, by using the power transmission network as the carrier. During the power assets restructure process, due to the flexibility of power demand is low, in most cases, it only required to separate power transmission from power supply, and encourage direct power purchase by major users, without against combination of power generation company and power supply company. Therefore, based on power transmission grid, some power distribution (power supply) enterprises and major users can establish transaction relations with power generation enterprises.

(2) Adopt multiple forms, ensure independent of dispatching and trading agencies, create preconditions for fair competition

The transient, network-based and security of power system determines the public property of power dispatching and trading agencies. To ensure stable operation of power system and objective and fair trading environment, almost all countries emphasized the objective of dispatching trading agencies. Among dispatch agency, transaction agency and power transmission enterprises, three combination approaches can be concluded as below: 1) independent system operation (ISO/RTO) mode, it refers to combination of power dispatching agency and power transaction agency and independent from power transmission enterprises. For example, in USA, Canada, Australia and Argentina, the independent ISO and RTO have no relations with market members, they only in charge of market transaction and power transmission to ensure openness and fairness of power transmission and ensure the independent of power system dispatching and fair competition in power market. 2) independent power transmission operation agency (TSO) mode, it refers to combination of power dispatching agency and power grid companies, independent from transaction agency. Most countries in EU adopt this mode, like north European power market, the power exchange agency is responsible with clearing service in financial and spot market; TSO is in charge of construction, maintenance and operation of power grid, as well as dispatching operation; 3) ndependent mode among dispatching, trading agencies and power transmission companies; in Brazil, dispatching agency and transaction agency was independent from power grid company, to establish power system operation agency (ONS) and power market transaction management agency (MAE). ONS is to coordinate the operation of power grid; and MAE is in charge of power transaction and clearing. In this way, qualified power generation companies can be ensured to connect to power grid, and users can select their preferred power supply company to obtain high quality power supply service.

(3) Promote, strengthen effective regulatory on power transmission and distribution cost and power tariffs

System innovation of power enterprises is to construct micro-basis of power market. If the competition only happens among state-owned enterprises, we would not call it real market competition. To reduce financial investment from government, change low efficiency of state-owned enterprises, UK and some Latin American countries actively promoted privatization of power assets during the power market reform, including power generation capital, power distribution capital and some power transmission capital. Up to now, all power capital in UK has been privatized; most power generation companies in Australia is private sectors, power transmission companies are state-owned, excluding Victoria state; the power transmission companies in Brazil and Argentina are state-owned, most power distribution and power generation companies are now sold to foreign and private enterprises. The characteristics of power industry determined its monopoly position in power transmission and distribution business. Independent power transmission and distribution tariff is the bridge between on-grid tariff and end-user tariff. The power transmission and distribution tariff directly determines the optimization allocation of limited resources and power transaction activities. Therefore, in addition to openness of power g rid and safety regulatory of power grid operation, effective regulatory on monopoly of power transmission and distribution should be strengthened, including regulatory in power transmission and distribution planning, investment, price, efficiency and services. Power transmission and distribution tariffs should be approved by government by using cost plus or price cap methods. As for independent power supply enterprises, retail rates should be determined with consideration of electricity tariff, operation fee of power distribution system, tax, cost-sharing fee and returns on investment in accordance with franchise contract. Meanwhile, to reduce overlap of subsidies, energy conservation, environmental protection, renewable energy development and other factors should be considered. In addition, power grid enterprises are no longer bearing any other social responsibilities.

(4) Bilateral transaction based, auxiliary with real-time balance, establish power wholesale market

A competitive wholesale power market is the main carrier of market reforms in power industry. According to the type of trade product, markets can be divided into electricity market, ancillary service market, market of power transmission right, capacity market and so on. Of which the electricity market is the essential part of the power wholesale market; ancillary service market and market of power transmission right is the useful complement. According to trade time, markets can be divided into real-time market, day-ahead market, and forward market; of which the real-time market and day-ahead market is normally known as power spot market, which is the core of power wholesale market; while forward market and future trading is the complementary product for power market to avoid risk. Classified by trade propriety, it can be divided into spot market and financial market. In the early stage of market construction, mandatory power pool model was adopted, such as UK, Australia and USA. With continuous reform in power industry, more and more countries encourage bilateral transaction, allow direct power purchase by major users; and the system dispatching operation agency is responsible for real-time balance. Through bidding or free negotiation, power producers can sign bilateral power purchase contract with power distribution companies or major users. The transaction contract could include power quantity, price, duration and transaction conditions. The dispatching agency will be responsible for scheduling, transaction and clearing after security check. In addition to bilateral contract, the rest generated power will be included in spot market. The price of spot market is determined according to load demand. Units are sorted by requested price, the price of spot market would be the same as the price of unit ranking at the bottom of the list which meets load demand. Power generation enterprise cannot meet power supply amount in the contract due to maintenance or accident, they can purchase electricity from the spot market to meet requirements of bilateral contract. The power market operation mechanism began to be mature.

(5) Strengthen construction of legal and supervision system; ensure orderly operation of competitive power market

A strengthened supervision system is the important safeguard of power market. During the reformation period, all countries clarified the exact role of which the government plays, and emphasized the necessary and importance of establishment of regulatory agency. Except the USA (Energy regulatory commission was established in 1977), UK, Spain, Australia, New Zealand, Brazil, Argentina, India and other countries established their independent energy or electricity regulatory agency since 1989. The EU requires membership countries to establish power and natural gas regulatory agency in the second Act; and in the third Act, it requires to unify the establishment mode of regulatory agency to found EU regulatory association, to promote unification of EU power markets. The main duty of foreign power regulatory agency is price system regulatory, based on the supervision goal of cost reduction and efficiency improvement: 1) to establish business license system, to clarify operation and quality standards directly related to power cost and tariff; 2) to strengthen market regulatory and ensure competitive in power market; 3) to establish power enterprises investment and performance standard; 4) to collect cost, benefit and performance data of power enterprises, to determine the power tariffs and operation performance; 5) to approve power tariffs to meet basic operation and investment needs of power enterprises; 6) to adopt uniform accounting system, and provide comparable cost data for pricing; 7) to be responsible for dispute resolution procedure between power enterprises and consumers; 8) to promote cost-effectiveness through management auditing; 9) to submit regulatory report to entity at higher level, and provide information and suggestions to related government; 10) to establish HR policy and procedure of power regulatory agency, recruite and train regulatory staff.

4.2.2 Enlightenment of foreign power market reform

There are a lot of successful experiences we can learn from power market reform in foreign countries, as well as classes to learn. Combined with current reform in power industry in the PRC, main enlightenment can be summarized as below:

(1) Prepare overall design and establish the legal system in the first place

To establish effective power market system, not only qualified market player and reasonable market structure is needed, but also legal system and market regulations. Therefore, the overall design should be prepared firstly, as well as legal and institutional safeguard. A successful reform relies on some basic institutional conditions, by clarifying duties of governmental departments and supervision agencies, obligations and responsibilities of each involved body, implementation plan of reform and supervision work, to improve the effectiveness of the reform.

(2) Ensure relative concentration of reform initiative

When the US and Australia promote the reform of power industry, they found that giving too much power to the state government while lack of authority of federal government is the main reason of unsmooth implementation of reform measures. The PRC should take advantage of our centralized system to ensure the initiative of central government, and ensure relative concentration of supervision authorities.

(3) Orderly promote the market construction of power industry

The power market should be promoted and improved steadily. Three power markets in northeast US were established based on 20 years of economic dispatching plus 10 years of evolution. Also, power markets in northern Europe and Australia have been through years of improvement. Besides, Latin American and Middle East European countries still have no effective power markets. We should not only learn their successful experience, but also summarize their failure lessons, combining with our own situation.

(4) Establish strongly regulatory agency

Firstly, to ensure the relative independent of regulatory agency, the government could not interference any decision making by the supervisor. The supervision goal should focus on the overall effectiveness of economy and maximum of social welfare. So that specific public and political objective of the government would not impact the efficiency of power industry; secondly, the responsibility of regulatory agency must be complete. Responsibilities such as market access, price regulation, power market regulation should be allocated to the regulatory agency.

4.3 Energy-efficient power generation scheduling and power market transaction

4.3.1 Relation between energy-efficient scheduling and market mechanism

The energy-efficient power generation scheduling was proposed to enhance energy utilization efficiency of power industry, promote energy conservation, reduce environmental pollution, promote energy and power structure adjustment, ensure safety and high-efficient operation of power system, and achieve sustainable development of power industry. The energy-efficient power generation scheduling is based on energy consumption and pollutant emission level, to replace auction mode which is based on price. Therefore, the price system seems loose its function in adjustment as economic signal.

However, the dispatch sorting based on coal consumption and grid integration auction based on principle of minimum cost is reform of power dispatch in the PRC. It is the transformation of power operation method and innovation of power scheduling regime.

Secondly, as for the goal, both focus on reduction in energy consumption and rational utilization of energy. Energy-efficient power generation scheduling pays more attention on reduction of pollutant emissions, while construction of market mechanism focuses on optimized allocation of power generation resources by using price signal.

Finally, as for the sustainable development of power industry, both ways target to sustainable development. Implementation of grid integration auction must stimulate more large units with low energy consumption connect to the grid through construction of power market mechanism and induce small thermal units with high energy consumption and high pollutant emissions to decommission, to eventually optimize resource allocation. Implementation of energy-efficient power generation scheduling is the improvement of supplement to power market operation regulation; it also speeds up the decommissioning of small thermal power units, promotes construction of regulations in power market, and helps to establish technical supporting system and power regulatory approach which is similar the power market.

We are searching the combination of energy-efficient scheduling and market mechanism to establish long-term mechanism of environmental-friendly energy-efficient scheduling in power industry, to fundamentally eliminate high energy consumption and high pollution and eventually achieve energy conservation goals. Therefore, how to promote power market construction through energy-efficient scheduling, or how to achieve energy-efficient environmental-protection and economic scheduling is a key point in the future research, as well as a realistic problem need to be solved urgently.

4.3.2 Particularity and feasibilities of the electricity market transaction

As a secondary energy, electricity has the common properties of ordinary products, for example, it has value and value in use; it can be purchased by currency like other products. However, as an important secondary energy and social necessity, electric power is different from other products with its special natures.

Firstly, production, circulation and consumption of electric power must be achieved only through a specific system—the power grid, thus the sale of power goods cannot be conducted through ordinary channels and methods fit to common commodities. It therefore requires power companies and the user must maintain close contact and friendship relationship with each other. Compared with that of ordinary goods, the establishment of the electricity market faces a certain degree of difficulty and specialty.

Secondly, electricity products cannot be stored in a large quantity with its characteristics of random demand, instantaneous transmission and also, the products own similar quality. All the features above restrict that the market transaction of electricity products should be produced in accordance with the expected sale situation and the products should be sold promptly after produced. Therefore, compared with other ordinary products, power market must establish the necessary technical support system.

Finally, due to the instantaneous balance between power supply and demand, market equilibrium price of electricity was random. Due to the restrict of units and the capacity of transmission lines, as well as that the demand elasticity is low, the electricity tariff jumps and shows a strong peak feature, so the electricity market must guard against price volatility, compared with other ordinary markets.

However, due to electricity is regarded as a highly standardized commercial product. There exist market demand-supply relations, price adjustment mechanism and market competition. It is qualified to establish power market-oriented mechanism.

4.3.3 Comparison and analysis of transaction mode

At present, in a competitive electricity market, the transaction modes mainly include the power pool mode, bilateral trading patterns, and the mixed forms of these above modes.

(1) Power pool mode

Power Pool model was designed on the basis of the theoretical starting point as below.

To establish electricity markets needs limited competition in the fields of power generation and sale of electricity. However, power energy strongly depend on the transmission network, the reliability and security of power system is strictly required, and decentralized power exchange may result in unsafe operation of power systems and the emergence of market giants, limited competition and proper monopoly-manner market could be designed for the power market.

The mandatory of power pool model only targets to physical transaction, which is electricity transaction. The financial transaction doesnot have to be conducted through power pool model, for example, financial contract signed between power generation company and power supply company or major users.

The power pool model can be divided into two modes:

1) **Non-competitive purchase mode:** bidding is carried out for sellers, power purchase right is even for buyers. In this mode, there would be no impact on power purchase tariff due to there is no power purchase bidding. The power pool model applied in UK at early stage, power market in California in US and few power markets in the PRC adopted this mode. Different from UK, the power market in California USA doesnot allow power buyers and sellers to sign financial transaction contract. The power market in Zhejiang is based on power pool model in UK, it allows differential price contract between power generation company and power grid company. The advantage is this mode is in line with current organization and management system of power market in the PRC; the disadvantage is high risks for power grid companies, and unfair competition among power generation companies due to the monopoly position of power grid companies.

2) **Competitive purchase mode:** in this mode, except bidding from sellers, power purchase bidding will also carried out among buyers. The final power tariff is determined by power demand and power purchase price provided by power supplier. New South Wales in Australia adopted this mode.

Compared with those two modes, the latter one can better reflect market rules that price fluctuation is impacted by market supply-demand relation, so that the power market is able to spontaneously and orderly operation. For the former mode, market risk is high; the market crisis might happen when supply cannot meet demand while electricity tariff is under regulatory. The crisis might happened to California power market.

(2) Bilateral transaction pattern

Bilateral transaction pattern refers that the power generation party or supplier signed electricity trading contract directly with demanders through consultation to determine the price and quantity. Bilateral trading patterns show the following characteristics:

——Buyers and sellers can form spontaneously the market forces of supply and demand for the optimal allocation of resources, to avoid interference from other non-market factors.

——To provide greater flexibility for main bodies of market. Power quality, quantity and price etc can be freely negotiated and ascertained by the parties for the transaction.

——Main bodies of the market can reduce operational risks through signing the combination of bilateral contracts, such as signing long-term power sales contracts or adding some terms on the price adjustment.

——More market players can be involved, which increased openness and competition of the power market. As the transaction between the generators and demanders could be conducted directly, monopoly factors in electricity industrial organization can be therefore effectively reduced and the competition in the market can be fully motivated.

Different with power pool model, this pattern is no longer mandatory power seller and buyer participate spot market. Most physical transaction is completed through negotiation. The trading quantity and trading price is no longer decided by centralized market. Under bilateral transaction pattern, system operator is only in charge of system balance, dispatching operation and power transmission management. NETA power market in UK adopted this model.

Compared with the power pool model, the bilateral transaction pattern is much closer to commodity market with certain advantages. First of all, under this model, market participants have more autonomy, emphasizing voluntary transaction based on negotiated price, rather than mandatory; secondly, this model provides more participation opportunity for buyers by price leverage; thirdly, the market risk of power pool entity can be shifted and weakened.

We should be aware that eventhough this model has its own advantages, and some proponents are promoting the bilateral transaction pattern, still we cannot ignore its disadvantages either. Firstly, the transaction cost for both sides is higher; secondly, the credit risk for both parties is relatively high; thirdly, the operation of a pure bilateral transaction has its difficulties. In a pure bilateral trading market, it is sometimes difficult to fully and timely control of market information, which would impact effectively operation of power market in a short-term. Besides, when obstruction or accident occurs, generally a central dispatching agency is required to relocate resources and coordinate operation status to achieve the healthy operation of the market. Finally, in the existing monopoly system, the operation control and management of power system is high concentrated. Transition to bilateral transaction mode from an old system is not conducive to a smooth transition.

(3) Mixed mode

Mixed mode, also known as net market mode, is a form of non-coercive power market.

This mode combines features of both above modes. There is power pool while also allowing for the financial transactions between the two parties and directly bilateral transactions. The both parties of provider and purchaser may independently choose to participate power pool transactions or for other forms of trading rather than follow a mandatory mode. Nordic electricity market falls into this pattern.

Mixed mode gives market participants maximum right to selection, and they can freely select any manner of purchasing and selling electricity based on comprehensive assessment towards cost, credit and market risk. Spot market operations of Power Pool provide a reference price to other forms of transactions. Under this model, the freedom and flexibility is greatly enhanced, combined advantages of two former models. It has become more respected in the international power market.

4.3.4 Diversification and multi-level of power transaction

An electricity market may include a variety of power trading co-existing and the modes of transactions will also diverse. Currently, transaction types and trading ways of the electricity market differentiate significantly. Trading modes in the electricity market, according to the different nature of the subject of the transaction, can be divided into two categories: physical electricity trading (also called physical trading) and the financial power transactions.

Physical transaction of electric power can be divided into electric energy trading, ancillary service transaction and power transmission right trading in accordance with transaction purpose. Classified by transaction duration, it can also be divided into forward transaction (bilateral contract transaction), day-ahead spot transaction, hour-ahead transaction (also known as daily trading or market regulation transaction) and real-time transaction (also known as balance transaction). Physical transaction of electric power is the primary trading of power products transaction.

Financial transaction includes futures trading, options trading, CFD constract for difference and financial power transmission right transaction. The financial transaction aims to avoiding market risk and hedging. After signed the contract, there would be no physical transaction, only financial settlement. Transaction types are shown as follow:

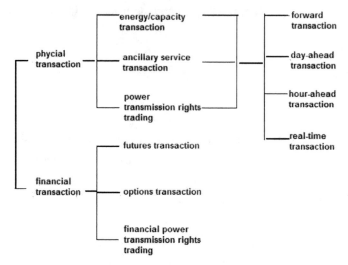

Figure 4-1   Power transaction mode

At present, China's power energy trading, with main stream of contracts transaction and spot trading as replenishment, can accept future transaction at proper time. Trading for the transmission and ancillary services can be conducted on the basis of qualified market condition after authorized by electricity regulatory mission[6].

4.3.5   Transaction mode based on energy conservation

Energy-efficient scheduling mainly targets on optimization of coal consumption; economic scheduling focuses on optimization on coal consumption cost, power purchase cost and grid network loss; environmental-protection scheduling emphasizes on optimization of pollutants emissions or pollutants discharge cost. A scheduling rule aiming to energy conservation and emission reduction should construct dispatching mode of pollutant emission cost or transaction of pollutant emission right based on market environment to achieve energy conservation, emission reduction and economic goals[7, 8].

During the actual operation, to achieve energy conservation, emission reduction and economic goals, attentions from multi-level shareholders should be considered to search a balance point to each side[9], to promote further reform. An effective power market is the essential which it doesnot exist in the PRC at present. The reality problem is how to design

---

6   "Primary Rules of Power Market Operation", by State Electricity Regulatory Commission, Oct 2005

7   Rongfen, et al, Overview of EEPGS study, Hydropower energy science, Febraury 2011.

8   Li Guanzhong et al, Auction mode with consideration of energy conservation under carbon trading mechanism, Power system automation, May 2011.

9   Zheng Bijian, Discussion on rise of China, 11th May 2011

energy-conservation mechanism based on market economic rules. At energy Saving point of view, it targets at coal consumption, economic mode emphasizes the systemic coal cost, power purchase costs and energy losses in the transmission system, and for the environmental protection point of view, pollutant emissions or emissions cost. Operation mode of energy saving should be decided considering the scheduling mode for pollutant emission quantity and emission cost or the scheduling mode based on emissions trading in the market environment, in order to achieve the comprehensively optimistic goal in terms of energy saving, emission reduction and economic using[10, 11]. The key question of reality is how to design the energy saving mechanism at the premise of government-led market economy, reflecting the market economy rules.

(1) The market mechanism to achieve energy conservation

According to economic theory, in a fully competitive electricity market, market participants tend to offer on short-term marginal cost, to achieve the objective of maximizing total social welfare and therefore obtain the best efficiency of the electricity market. Moreover, through the reasonable market institutional arrangements and scientific price mechanism, the various market players could be stimulated to actively consider the cost of various energy resources consumption, prompting market players to strengthen the business management, innovative technology and actively strive to reduce energy resources consumption including coal during construction and operation period and all the above will be reflected by the price level. The overall energy consumption of lower-price quotation is relatively low, with strong market competitiveness, and ultimately makes the result of internal unity of market competition and emissions-reduction.

In the transmission link, measures were taken to reduce energy losses in the transmission system according to the market rules. For example, by means of amendments to offer discount net loss, under the same conditions, then the unit with low transmission loss will enjoy priority scheduling. We also introduced node marginal price mechanism so that spot pricing could reflect the transmission network congestion and loss information in different locations. Besides, the reasonable market mechanisms can also contribute to the transmission side to reduce energy losses in the transmission system.

Market mechanisms were adopted by sales party, which helps optimize the user's power consumption method, while the optimized allocation of resources consumption can

10   Rong Fen, etc, General Guidelines of Energy-Saving Power Generation Scheduling, "Hydropower Energy Science", January, 2011

11   Li Xianzhong, etc, "Bidding Pricing Mode under the Framework of Carbon Trading Considering Energy-saving and Emission Reduction", Automation of Energy Market, May, 2011

also help achieve energy-saving goals. For instance, the user response to real-time electricity market prices and therefore automatically reduce electricity price peak load to achieve the energy saving target. In addition, through price response, the user autonomously conduct the load shifting to ease the power plant expansion, lower the system investment pressure to increase in peaking power supply , to reduce consumption of energy and resources. Therefore, scientific decision-making power through the market mechanism the optimal allocation of resources to achieve the power side of the energy consumption.

In summary, the market mechanism can promote multiple aspects of power system operation for all-aspects energy consumption, to achieve an integrated energy-saving target.

(2) Implementation of macroeconomic policy to promote energy conservation and emission reduction

If we do not consider the external negative power generation, the market mechanism failing in the promotion of emission reduction is inevitable. The environmental costs can also be considered as a resource, i.e. environmental resource, through the market mechanisms design which can facilitate the internalization of external cost, we could achieve the goal of optimal allocation of resources, emission reduction and promote environmental conservation purposes. Thus, by the government's macro-control means we can resolve the problem of generating negative externalities and improve the market mechanism to achieve emission reduction.

Foreign experiences show that through the market mechanisms design which can facilitate the internalization of external cost, we can use the market competition method to allocate power generation and environmental resources to achieve emission reduction. For example, some EU countries have started to impose high carbon tax levied on burning of fossil fuels. Cost for purchasing CO_2 emission right contains a certain proportion of power generation cost. Therefore, integrating the external environmental cost into internal cost could achieve optimization of power generation resources and environmental resources through market competition method to thereby achieve energy conservation.

(3) Energy conservation: transaction patterns can learn from

1) Korean mode of cost-based power pool

At present, the total installed capacity in Korea reached to 64.5 million kW, of which capacity of nuclear, coal-fired, gas-fired units reached to 18 million kW; oil-fired and hydropower units reached to 4 million kW respectively; voltage of power transmission line is classified to 154KV, 345KV and 765KV. Before reform, the KEPCO is the main body of power industry in Korea; its six power plants generated 94% of total electricity in Korea. In addition, KEPCO also monopoly of all power transmission and distribution business in Korea.

In 1999, the Korea Trade and Industry Department of Energy announced "The Reform Program of Electric Power Industry" in accordance with the direction of market-oriented reforms and implemented it in four steps. In 2001, policy of "separation of power plants from power grid" was implemented. Except for nuclear power, six power generation companies will be gradually reorganized as a subsidiary of KEPCO or private companies and established a "cost-based power pool". Also independent KPX will be established and KEPCO will be only responsible of power transmission, distribution and electricity sale services.

KPX is mainly responsible for power market operation, power system operation and power system planning; it is a non-profit organization. The cost-based power system operated by KPX has 7 members, including KEPCO and other six power generation companies, with total capacity of 44.39 million kW accounting for 82% of total installed capacity. Since 2003, the power system was opened to large account users of 50,000KVA. By the end of 2004, the total number of members reached to 56, with capacity of 58.94 million kW, accounting for 92% of total capacity. As a non-profit organization, the duties in market operation of KPX include auction, measurement, market regulatory, information release and dispute resolution.

As it is cost-based power system, except power plants under PPA contract, all power generation companies and power selling companies must complete transaction based on the power market. Few qualified major users can directly conduct transaction with power generation companies. The power plants submit cost change of each unit to power system; the cost evaluation commission will determine change cost of each unit monthly, and capacity cost (construction cost and fixed cost) of each unit annually; KPX establish scheduling planning according to the cost of power generation unit; system marginal price reflects the actual cost (including setting up cost, idle load cost and slight increase cost) of marginal cost; capacity cost will be charged for on-use units (no matter it is dispatched or not); the power purchasing price in the market is based on system marginal price and capacity price of generation unit.

The cost-based power system in Korea has its own characteristics. The power generation units are sorted based on change cost. It is similar with energy-efficient power generation scheduling in energy conservation. Its market organization system, transaction mechanism and price formation mechanism is shown as Figure 4-2.

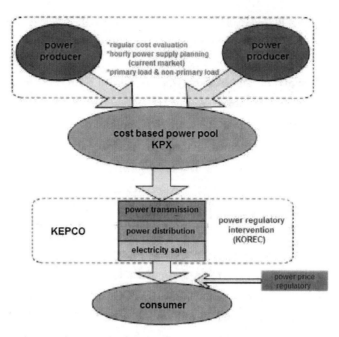

Figure 4-2   Market organizational system of cost-based power system in Korea

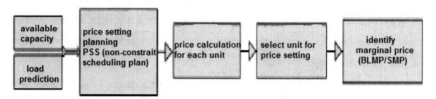

Figure 4-3   Marginal price identification procedure of cost-based power pool

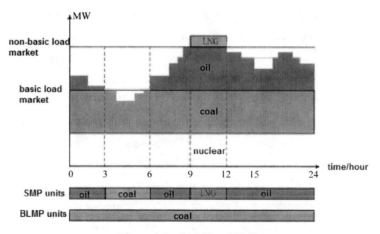

Figure 4-4   BLMP and SMP

The benefits of such transactions is to ensure the capacity cost and investment return while introducing the competition mechanism to ensure stable prices. This could both promote energy-saving and emission reduction and ensure the stability of the system. The system could realize the objective generation cost but also to ensure fairness and justice of dispatching and transaction. In addition, it won't produce stranded costs, and prevent from market interference.

The drawbacks are similar to the British Library in early days. On one hand, purchase was conducted in single format and KEPCO remains a dominant market position. On the other hand, there isn't price transmission mechanism and can't be regarded as the true electricity market. However, in the early period after the electricity market was established, especially under the system with transmission and allocation integrated, as a transitional form of the electricity market, it could achieve basically a combination of energy-saving and emission reduction as well as market mechanisms.

2) Solution recommended by World Bank

In May 2010, the World Bank pointed out in a policy suggestion report of "To Improve the Efficiency of Chinese Power Generation Scheduling" that in China, to improve the efficiency of power generation dispatching is an important measure to achieve substantial savings and emission reduction of coal resources. The proposal was titled "Energy-Efficient Power Generation Scheduling Method", which is an important innovation stepping forward for low-carbon power. To solve the issue of financial compensation, we may take the following three alternatives:

First one is financial contract. With this method, the operation hour will be transferred to energy through financial contract, similar with power generation right transaction. The power generation company sells its power generation right to provincial power grid company according to power on-grid tariff. Without balanced power tariff, power generation company can clearly define the energy difference generated from contracted energy and efficient scheduling. The key point is to determine rational standard and pricing rule of contracted energy.

The second one is administrative financial compensation. Extra income from high-efficient power generation companies based on administrative procedures and pricing rules should be compensated to power plants with low efficiency and having negative financial impacts. The key point is to define annual power generation baseline of units and difference between on-grid tariff and baseline.

The third one is two-settlement power pricing. Power on-grid tariff is divided into capacity tariff and energy tariff. Capacity tariff is based on constant cost of marginal coal-fired units to ensure necessity degree of security backup; energy tariff is based on constant cost of coal-fired units. The advantage of this method is that it can be well connected to power market and can easily create a competitive capacity market and marginal electricity auction market.

Common practice of international community is to minimize emissions at the lowest cost dispatch, the main differences between which and energy-efficient power generation scheduling show in two aspects, first is taking the pollution emission factors into accounts and the second is the stressed factor is the lowest power generation cost rather than least fuel consumption, so that it's more likely to dovetail with the market mechanism.

4.3 Regulatory policy construction of energy-efficient power generation scheduling

Under current economic development stage in the PRC, The energy-efficient power generation scheduling is considered as the temporary scheduling mode during the transition of market economy from planned economy. Along with the establishment of energy conservation pricing system, the new scheduling rule would be developed to a higher market operation stage. Barriers caused by coal consumption measurement should be resolved through supervision system urgently. But in a long term, it requires institutional enforcement. The State Electricity Regulatory Commission and its branches should be responsible the overall regulation of the implementation of new scheduling rule, power generation right trading and direct power purchase by major users. Besides, to ensure energy conservation in a long term, supervisions on technical supporting system design, coal consumption online monitoring system, operation of desulfurization units, public release of units ranking list and timely clearing of electricity sale.

4.3.1 Regulatory on information dissemination

The rule of energy-efficient power generation scheduling is in accordance with "Information Dissemination Approach of EEPGS (trial), Electricity regulatory market (2008) No.13". It classified information dissemination policies according to different objectives. The state power scheduling communication center is responsible for updating release information from the website.

The energy-efficient scheduling is different from traditional scheduling. It includes coal consumption and units ranking information. The original publisher of above information is market competitor who expects more benefits. The authenticity of the information directly determines the healthy operation of power market. The new rule does not have administrative constraints as planning dispatching or contract constraints as market-oriented dispatching. The stability of the new rule can only be achieved under triangle stable: market risk, power regulatory and information system technology. Therefore, the establishment of effective regulatory policy on information dissemination is essential. Main contents include: Firstly, to clarify the significance of scheduling information, information content released by electricity regulatory agency, related governmental departments, electricity scheduling agency and power plants; Secondly, to clarify the release method, definition of release method, time and objects; Thirdly, to effectively reflect the function of electricity regulatory agency and provincial governmental departments. Detailed information includes:

I. Annual (monthly) load forecasting

(1) total power generation amount and total power demands;

(2) maximum and minimum load;

(3) monthly average power consumption load rate.

II. Annual (monthly) units combination program

(1) units ranking list;

(2) installed capacity and adjustable capacity of each unit;

(3) monthly maximum load, maximum average load, monthly power demanding forecast at provincial (regional) level;

(4) annual (monthly) maintenance planning for units and power transmission equipments;

(5) operation planning of reservoir in each hydropower plants;

(6) operational planning of power generating equipments;

(7) decommission planning of power generating equipments;

(8) annual (monthly) planned power generation hour for adjustable and unadjustable renewable power generation units and variety factors of next year (month);

(9) annual (monthly) planned power generation hour for nuclear power generation units, coal-fired cogeneration units, residual gas, residual pressure, coal gangue, washed coal, coal bed methan based units, natural gas, coal gasification based units; and variety factors of next year (month);

(10) to be scheduled units for next year (month).

III. Other information

(1) energy consumption level of generating units;

(2) energy saving effects.

To clarify the rewards and punishment by establishing morality risk mechanism. Pre-evaluation and post-evaluation should be carried out for coal consumption and other key information to ensure the implementation and application of energy-efficient scheduling.

4.3.2 Regulatory on coal consumption index

New challenges have been come up along with the implementation of energy-efficient power generation scheduling. The new scheduling rule need to be further optimized aiming operation characteristics of power grid. For example, generation units are classified according to their coal consumption, which requires better method for higher accuracy. Also, in some areas, the right of on-grid scheduling is not unified due to some users have their own power plant. In this case, the scheduling approach should be clarified. To achieve the goal of energy conservation and ensure timely completion of the goal, optimization mode targeted on minimum energy consumption and minimum power purchase fee should be carried out.

4.3.3 Regulatory on priority of renewable resource and its quotas

The State Electricity Regulatory Commission and its branches should immediately launch the regulatory content of energy–efficient power generation scheduling, and apply the regulatory on grid-connected renewable energy generation and its scheduled quotas. The scheduled quotas of renewable energy should be ensured through multi-channels, including main grid network, micro grid and dispersed energy supply system. Details include:

Firstly, to provide favorable renewable power grid planning, investment, approval and regulatory policy to promote frastructural construction for energy-conservation power grid; secondly, to provide policy protection for renewable energy and substituted power generation, and to clear on-grid technical barriers for new energy; thirdly, facing with great power demand when extremely weather happened, the extension function of energy-efficient power generation scheduling is expected to integrate all new energy based power generation to avoid impacts of serve life; fourthly, to strict regulate the energy efficiency index of power plants and release quotas of renewable energy based power generation.

4.3.4 Regulatory on "Large Substitutes Small" program and power generation right trading

The regulatory content on the LSS program is fair trade of scheduled quotas. For example, at the beginning of implementation of energy conservation policy, "scheduled generation quota trading" policy can be carried out for small power plants. This policy aims to giving more power generation quotas to large efficient units through purchasing scheduled quotas from small inefficient units and decommissioned units. It provides a fair trading environment of power generation rights for power plants. It also establishes a public discussion area for renewable energy based power generation and users complaining.

4.3.5 Regulatory on direct power purchase by major users

Normally, when the power transmission line is not blocked, the grid company would ignore the power purchasing price, but care about the power transmission fee. But when the power transmission line is blocked, the power transmission cost would be increased. The dispatching or power transmission company might cancel the direct power purchase agreement with major users. Therefore, the main point in this area is regulatory on power transmission cost. When major users directly purchase electricity in day-ahead market and contract market, dispatching officials might not fully implement the day-ahead trading contract due to actual demand and coal consumption ranking. The differential part would be compensated with balancing market. Therefore, regulatory should be carried out for all steps above.

5. Interaction between Power Price and Energy-Efficient Power Generation Scheduling

5.1 Constitution of electricity tariff system in the PRC

The electricity tariff reform in the PRC focuses on: establishment of grid-connected power price mechanism; power transmission and distribution price mechanism which is helpful to promote healthy development of power grid; joint price system of gird-connected power price and sale price; optimization of sale price structure; and direct purchasing of major users. The final task of tariff reform includes: power price would be divided into: grid-connected power price, power transmission price, power distribution price and terminal sale price; power generation and power sale market should be formed by market competition; power transmission and distribution price should be controlled by the government. Logical relationship of price system is shown in Figure 5-1.

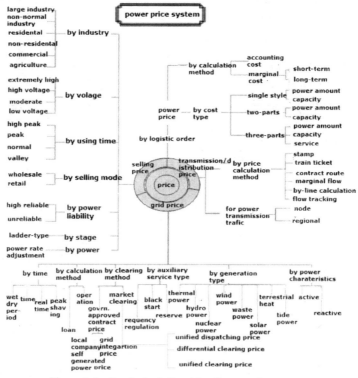

Figure 5-1　Logical relationship of power rice systems

5.1.1 Power on-grid tariff

The "grid-connected electricity tariff" refers to the clearing tariff of grid-connected electricity when power purchaser buys from the power plant.

(1) Time-of-day tariff structure

The on-grid electricity tariff can be divided into flow high water/low water tariffs, peak/valley tariffs and real-time tariff.

(2) Power on-grid tariff by clearing approach

The on-grid power price can also be divided into government-approved price and market clearing price according to different clearing approaches. Government-approved tariff includes auto-supply power tariff, grid-connected price and unified scheduled price; market clearing price includes capacity charge and energy charge. Capacity price is based on available electricity of units; energy price is based on on-grid electricity amount. Energy price can be further divided into differential clearing tariff and unified clearing price.

(3) Power on-grid price by power generation type

Power on-grid price can be classified according to power generation type, including thermal power, nuclear power, hydropower, wind power or waste based power, solar power, tide power, geothermal power and so on.

(4) Power on-grid tariff by power property

Classified by power property, the on-grid power price can be divided into active and reactive power price. If user power is too low, bigger current is required to protection normal operation of electrical machines. The increasing current of power transmission line would increase heat loss, voltage loss, reduce the service life of electrical equipments and increase energy loss of the equipments thereby increasing user's electricity cost. Reactive power price is proposed for reactive compensation, to increase power factor and maintain its equilibrium, reduce reactive loss, improve power factor of power grid and equipment power, to improve social benefits and benefits for electricity suppliers and users.

(5) Classified by ancillary service

It can be divided into frequency regulation price, reserve power price and black start power price.

(6) Classified by policy regulatory

It can be divided into operation period price, capital & interest price, benchmark price and coal-electricity linkage price.

1) Capital & interest power price

Before 1998, fund-raising policy was carried out in power industry. On-grid power price for power plants which were not financial invested by government (independent financing, sino-foreign joint venture), should be based on their loan and benefit situation. The goal is to guarantee that investors would pay back the loan and gain profit in loan period (normally 10 years).

2) Operation period power price

In 1998, power price in operation period is issued, and "capital & interest power price" was improved to extend the repayment period. It is basically based on average cost in accordance with revenue level and social average cost during operation period to constraint investment cost and operation cost and reduce power on-grid price.

3) Benchmark power price

In April 2004, the NDRC determined to unify power price of new operation power plant, which is benchmark power price. Price management department determines the same level of on-grid power price of same type units in accordance with operation duration of power generation project, in line with principle of "reasonable cost and reasonable revenue". Annual on-grid power amount during operation period is analyzed according to units' situated areas to identify on-grid benchmark power price and carry out bidding. The hydropower/thermal power on-grid price of each area is determined by different areas, power price of newly constructed power generation projects is determined by regional/provincial average cost. The benchmark power price of each area is determined based on unit average cost, coal cost and coal consumption. Benchmark power price can provide power pricing policy and power price signals for investors. The government could encourage or restrict power investment, adjust power investment structure, promote resource optimization allocation and reasonable capital flow by establishing and adjusting benchmark power price. The power producer can also calculate their benchmark production cost according to benchmark power price.

4) Coal-electricity linkage price

In 2004, the NDRC defined, based on integrated price of coal used for power generation, a linkage period is 6 months or above, if the adjustment rate of average coal price reached to 5% or above than precious period, on-grid power price should be relatively adjusted. The electricity sale price should be linked with on-grid power price based on principle of maintaining power transmission and distribution power price.

5) Integration auction power price

In 2002, the State Council issued power system reform program of "separation of power plant from power grid, integration auction". The power generation assets of State Power Grid are allowed to sale. Five state-owned power generation enterprises (Huaneng,

Datang, Guodian, Huadian and Central Power Investment), National Power Grid and China Southern Power Grid companies are established through merger and reconstruction. Few years later, northeast and east China power markets carried out integration auction power price pilots. Power generation settlement price is divided into two-settlement settlement and one-part settlement power price. Currently, the pilot work has been completed; power market still does not use market mechanism to allocate power resources; allocation of power generation quotas is still arranged through planning mechanism; even-load utilization hour is allocated by power dispatch center based on installed capacity of power plants. The on-grid power price of each power generation enterprise is government-directed power price mainly based on power generation cost.

6) Direct purchased power price of by major users

Since 2004, some regions have started to carry out program of "direct power purchase by major users". Large account users can directly purchase electricity from power producers only if their electric power and power consumption amount reached to a certain level. Power grid enterprises charge certain fees as on-grid power price through provision of power transmission and power distribution services. The first direct power purchase pilot happened between Jilin Longhua Thermal Power Plant and Jilin Carbon Corporate. Supported by the NDRC and State Electricity Regulatory Commission, those two enterprises signed the direct power purchase agreement. The agreement defined, factory price of Longhua is CNY 0.37/kW, on-grid price is CNY 0.139/kW, plus CNY 0.02/kW for Three Gorges construction fund and rural power grid transformation fund, thereby the final power purchase price by Jilin Carbon is CNY 0.41/kW.

5.1.2   Power transmission and distribution price

Power transmission tariff refers to tariffs charged by power grid enterprises for the provision of electricity transmission services through power transmission network. Power distribution tariff refers to tariffs charged by power grid enterprises for the provision of electricity distribution services via power distribution network. Establishment of power transmission and distribution tariffs should be in line with following principles: 1) promote social allocation efficiency. Without external constraint for power transmission and distribution enterprises, they would establish the market price but not accept the price. They might establish monopoly price to convert consumer surplus into consumer surplus, to uneven allocate efficiency. It requires that government implement price regulatory to promote social distribution rate. 2) promote efficiency and production rate. Through price regulatory policy to inspire optimization of production and operation management, to continuously carry out technical innovation and management innovation to achieve maximum production efficiency. 3) protect development potential of enterprises. Power transmission and distribution industry is characterized as big investment and long period of returns on investment. To ensure reliable power transmission and power supply, capacity of large-scale investment of enterprises should be considered during preparation of price regulatory policy.

(1) Power transmission and distribution price

The transmission and distribution grids in the PRC mainly include regional power grid and provincial power grid, the transmission and distribution tariffs can be divided into:

1) grid connection tariff of regional grids, transmission tariff for specific projects;

2) access system tariff of regional grids, regional grid transmission tariff;

3) access system tariff of provincial grids, provincial grid transmission and distribution tariffs.

Market clearing power price includes capacity price and energy price. Capacity price is based on maximum power transmission capacity. Energy price is calculated by transmitted/distributed power amount.

(2) Power transmission and distribution price

Generally, the regional power transmission grid provides two types of services. One is provision of system service for power plants to access regional power grid; power price for this part is access price paid by power plant, the provincial power grid to recovery through power purchase price. The other one is service provision for provincial power grids. Due to the difference of power grid structure, power source and load ratio, services provided by regional power transmission grid can be divided into two parts, firstly, mainly for power transmission as power transmission channel, to balance power supply and demand within regional power grid; secondly, regional power grid is used for both cross-provincial power transmission and backup function. For the first type of regional power grid, the power transmission price can be unified in the region, paid by provincial power grids; for the second type of regional power grid, power transmission price can be determined by grid security and power transmission functions. Price for security function is paid by provincial power grid, integrated to provincial power transmission and distribution price; price for power transmission function can be integrated to power purchase fee of power grid and recovered by electricity sale.

Regional power grid is linked through networking project. Meanwhile, specific power transmission project can also achieve cross-regional power transmission through point-grid or grid-point ways. The first one uses networking price paid by both sides. Regional power transmission price paid by provincial power grid; then recovered as power transmission and distribution price by provincial power grid. Power transmission price for specific projects is applied for latter one; normally it is included into on-grid power price paid by power grid that purchases power then recovered by electricity sale.

Public network is the core power transmission and distribution grid shared by all users. It is the middle linkage of power transmission. The fee can be shared by utilization intensity of users.

As for provincial power grid, most projects are 500kV power transmission projects, but also 220kV projects and power distribution projects. Provincial power grid companies mainly provide two kinds of services, one is providing access systems for power plants to access provincial power grid; the access fee is paid by power plant; the other one is providing power transmission and distribution service; the power transmission and distribution fee is paid by end-users and wholesalers. In addition to access price, power transmission price and power distribution price to recover provincial power grid cost, the provincial power grid also uses power transmission and distribution fee to pay transmission fee to regional power grid company.

(3) Power transmission and distribution tariff regulatory

The NDRC proposed two stages of power transmission and distribution reform in power pricing program in 2003. The first stage is, power transmission and distribution price is determined by difference between current actual average electricity sale price and average power geenration price; the second stage is, cost plus mode. In the power price reform program issued by the State Council in 2005, it is proposed to separate power transmission from power distribution business; and power transmission and distribution price applies cost plus mode. According to this mode, the allowed revenue for power grid companies is composed by allowed cost and allowed returns on investment. The allowed cost includes depreciation and operation and maintenance cost; returns on investment is determined by reasonable rate of return and effective asset. Reasonable rate of return is identified by considering debt ratio, bank rate and social average return to investment value. Depreciation fee is identified by effective asset and pricing depreciation rate; operation and maintenance fee should consider social average cost and effective asset, it can regarded as effective asset multiple average operation and maintenance rate of power transmission and distribution network.

According to the price regulatory theory in natural monopoly industry and power transmission/distribution charateristics, the power transmission and distribution price should be approved based on, one is rate of return, two is performance.

(4) Regulatory module based on return on investment

1) Cost recovery regulatory

Cost plus regulatory is one of average cost pricing methods, based on effective cost, allowing power grid enterprises obtain reasonable returns on investment.

$$RR_i = RB_i \times RoR_i + C_i$$

Where, RR_i 为 total income of year i; RB_i is the baseline of return on investment of enterprises, which means effective captical; RoR_i is allowed ruturns on investment; C_i is cost of enterprises, including depreciation expense, maintenance fee and tax.

Average power transmission tariff in year i is P_i:

$$P_i = RR_i / Q_i$$

Where, Q_i is transmission power volume in year i.

Different price can be reformed due to annual revenue demand, such as two-settlement power price, power transmission/distribution price by voltage level, nodal power transmission/distribution price and etc. The key point for this mode is how to regulatory all factors by government. Regulatory methods include cost regulatory, investment regulatory and revenue regulatory. At different development stage of power grid, different regulaotry strategy can be applied. For example, for developing power grid, relatively loose regulatory policy can be used to attract investment; for developed power grid, incentives can be established to reduce cost and improve efficiency.

2) Asset regulatory

Asset regulaotry mainly aims to identify asset baseline to calculate returns on invesmtnet, which is effective asset. Three methods can be used to evaluate effective asset: historical cost estimation, optimized deprived value and market value estimation.

Following historical value used to calculate returns on investment baseline:

Baseline of returns on investment includes: current asset (cash, inventory, repaid expenses and etc,.), net fixed asset, intangible asset (including land, land use right, patents and etc,.). in Norway, calculation of returns on investment of power grid companies is based on legar asset value of power grid plus cash (1% of fixed asset).

Optimized deprived value method to calculate returns on investment baseline:

The small value between optimized deprived value (deduct depreciation) and commercial value (economic value) is used in this method. Optimized deprived value refers to cost to the firm if it is deprived of its assets. It aims to re-evaluate the effective asset based on current value. The economic value of asset is the big value between net value to continuously use these assets and revenue to disposal of assets. When optimized deprived method is used to calculate returns on investment baseline, asset base includes both the value of fixed assets, and current capital (cash and all types of deposits, etc.) to meet requirements of short-term operation.

Market value used to calculate returns on investment base:

$$\text{Market value} = \text{share price} \times \text{shares quantity}$$

Normally, investors buy shares of power grid enterprises only through stock market with market prices. As for investors, calculation of returns on investment is based on purchase price. Therefore, regulatory agency should identify return on investment base according to market value.

3) Revenue regulatory

Revenue regulatory mainly aims to identify the asset benefit rate. There are a lot of methods used to identify returns on assets. But weighted average cost of assets is the most widely used one.

Formula for after-tax weight average cost of assets:

$$WACC = \text{Re}\ \times \frac{E}{V} + Rd \times \frac{D}{V}$$

Where, Re: cost rate of equity capital; Rd: cost rate of debt capital; E: legar value formed by equity; D: legar value formed by debt; V: legar value formed by equity and debt.

Cost rate estimation of equity capital:

Equity capital cost rate is returns on profit to compensate the investment of investor. There are a lot of methods to estimate equity capital rate, of which capital asset pricing model is a widely accepted tool, calculated as:

$$\text{Re} = Rf + \beta e \times (Rm - Rf)$$

Where, Rf is risk-free return rate, standarized by medium-/long-term bond issued by national development bank. In Australia, power grid regulatory agency calculate the risk-free rate according to daily average value of 10-year bond market price in recent 40 days of federal government.

$(Rm - Rf)$ is equity risk rate, equals difference between returns on market and risk-free return rate. It is normally obtained through analysis of difference between historical return on equity and risk-free rate.

β is the ratio of system risk of investors and whole market risk. $B > 1$, indicates the system risk faced by the company is higher than market average risk; it is high-risk investment; when $\beta = 1$, it means the system risk of this company and the system risk of market is the same, even risk can gain even returns; if $\beta < 1$, it indicates the systematic risk of investor is low, return rate would below average too.

β value includes two kinds: equity β and asset β, their relation can be simplified as:

$$\beta e = \beta a \times (1 + D/E)$$

where, βe is equity β, and βa is asset β value.

In UK, the regional power company has its asset β ranging between 0.6~1.0, with average of 0.8; in USA, power company has its asset value around 0.5; and it ranges between 0.4~0.5 for power company in Australia.

Estimation of debt capital cost rate:

Floating interest rate is used in foreign commercial bank, and loan rate as commercial secret would not be penned to the public; therefore, when the government or regulatory agency determines the debt capital cost rate, it can only be calculated as risk-free rate plus marginal return rate, as:

$$Rd = Rf + Dm$$

Where, Dm is debt marginal return rate, normally it ranges between 0.3~1.7% in UK and Australia.

(5) Performance based regulatory mode

Performance based regulatory mode is based on return rate regulatory. It is used to calculate price or revenue cap for the next 3-5 years within regulatory period. Annual adjustment would be carried out according to specific rules during the regulatory period.

1) Price-cap regulation

The price-cap is normally happened at the primary period of price-cap regulatory. The starting price is approved by regulatory agency, then adjusted with consideration of RPI or CPI during regulatory period minus the efficiency coefficient X.

Related agency in UK uses price-cap method to identify and adjust the power grid utilization fee. During the regulation period, maximum average price per kW should be calculated annually, and adjusted by a coefficient based on previous actual implementation situation. When the actual revenue of last year is higher or lower than approved revenue, the revenue of next year would be modified by increase or decrease a certain value. Annual maximum average price per kW is calculated as:

$$Mt = [1 + (RPIt - Xg)/100] \times Pt - 1 \times Gt - Kt$$

Where, Mt: maximum average price per kW in year t; RPIt: retail price increase index; Xg: efficiency coefficient; Pt-1: price in year t-1; Gt: maximum demand related index in recent years; Kt: correction factor of year t.

2) Revenue-cap regulation

Revenue-cap regulation refers to maximum allowed revenue of monopoly business determined by regulatory department. When we calculate maximum allowed revenue of monopoly business, firstly we calculate non-adjusted maximum allowed revenue; then, adjust maximum allowed revenue by predicted commodity price and CPI-X. Normally, allowed maximum revenue is calculated first then related price of monopoly business.

Revenue-cap regulation is calculated as:

$$MAR_t = MAR_{t-1} \times [1 + (PI_t - X)/100] + Z_t$$

Where, MAR: maximum allowed revenue; PI: price index; X: efficiency factor; Z: un-predicted revenue adjustment which cannot be controlled by regulatory officer; t: regulatory year.

5.1.3 Electricity sale price

Electricity sale price refers to end-users electricity price charged by power grid enterprises.

(1) Classified by voltage level

Divided by voltage level, the electricity price includes UHV tariff, high-voltage tariff, medium-voltage tariff and low-voltage tariff.

Electricity flow is from high-voltage level to low-voltage level. The end-user electricity price is classified by voltage level. The voltage level of power transmission grid is divided into 1000 KV, 800 KV, 500 KV, 330 KV and 220 KV. 1000KV and 800 KV are classified as UHV.

(2) Classified by time-of-use

Divided by time-of-use, the electricity price includes critical peak price, peak price, valley price and normal price. Installed capacity of power system is normally determined by load at peak period. The increase of load at valley does not need to increase installed capacity. Therefore, electricity consumption during peak period should be responsible for marginal cost of peak capacity and marginal cost of peak energy. Electricity utilization at valley should only be in charge of marginal cost of valley energy.

(3) Classified by users

The electricity tariffs can be divided according to end-users and use capacity, including residential electricity, electricity used in non-industrial production, electricity used in normal industry, electricity used for large-scale industry, electricity used in agricultural production or for other purposes.

(4) Classified by reliability of power supply

According to different requirements on power supply reliability, high-reliability power price and interchange power price is gradually implemented to reasonable adjust power demands. High-reliability power price is based on current power price with appropriate increase. The interchange power price is appropriate lower than current power price. To increase power supply reliability, backup capacity and backup circuits would be increased; thereby the power supply cost would increase. Therefore, users require high reliability power supply should undertake higher power price. When the power supply is insufficient during peak period, power sellers can temporarily reduce or interrupt power supply. Lower power price can be applied for electricity users under above situation. The lower interchange power price can be regarded as economic compensation for users who contribute to system load balance.

(5) Classified by power utilization volume

According the electricity volume by different users, the power price can be charged as ladder-like power price. In this price policy, users who consume more volume should be charged higher power price to embody fair principle. Due to ladder-like power price, users with high income will have to consider the high cost of electricity to reduce electricity waste and enhance energy utilization efficiency.

(6) Classified by cost recovery form

According to cost recovery form, the power price can be divided into one-settlement power price, two-settlement power price and three-settlement power price.

One-settlement power price is based on traded power energy or capacity, including one-settlement energy price and one-settlement capacity price. Two-settlement power price is based on capacity and energy, consisting capacity price and energy price; of which capacity power price is based on rated capacity accessed to system, mainly used for fixed cost recovery; energy power price is based on actual traded power volume, mainly used for changeable cost recovery. Sale capacity price is based on maximum demand of users, it is also known as basic power price; sale energy price is based on volume measured by meter. Three-settlement power price is based on capacity power price and energy power price, plus service based power price, mainly used by user side.

(7) Classified by sale mode

Based on sale mode, the electricity price can be divided into direct supply power price, wholesale power price and retail power price. Wholesale power price refers to power price sold by provincial or above power grid to county or below level users. It is based on a certain discount, around 20~25%, 28% in some areas.

5.2 Function analysis of power price system to energy conservation

It is needed to establish a price mechanism which can reflect resource scarcity situation, market supply-demand situation, environmental pollution and ecological damage costs, to effectively guide optimal allocation of resources. Grid power price system should be gradually changed into market oriented price system, to reflect market demand-supply relationship, to provide economic signal for investment, power transmission service and utilization of power transmission resource, to further improve linkage mechanism of grid-connected power price and sale price, to encourage clean energy development.

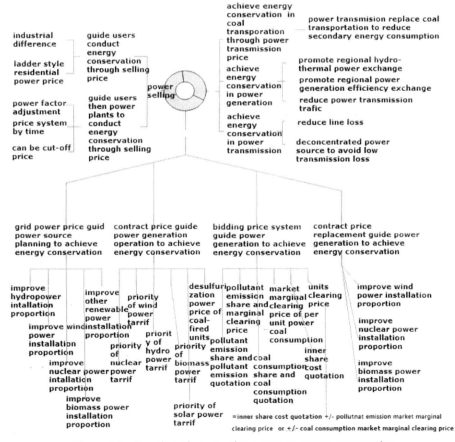

Figure 5-2   Function of power price system to energy conservation

5.2.1 Benefits of energy conservation from power generation price

In addition to reflect the cost of power generation, the power price should also reflect energy consumption cost, energy scarcity, environmental damage cost and supply-demand relations. For example, acid rain is caused by carbon dioxide emission during power generation. The power generation company should pay the use of environment resources, in order to conduct energy conservation policy.

(1) Power pricing policy for renewable energy

The "Interim Measure on Allocation of Income from Surcharges on Renewable Energy Power Prices" and "Interim Measure on Allocation of Income from Surcharges on Renewable Energy Power Prices" clarified related power price policy of renewable resources. The price difference between renewable energy and grid-connected power price of local desulfurized coal-fired generator would be allocated to selling power capacity at provincial and above level. Grid power price of wind power uses governmental-directed price, and power price standard is determined by bidding price from related price department of State Council. The on-grid tariff for the biomass powered plant shall be determined according to the benchmark tariff issued by the relevant authorities for price control. The on-grid tariff over the first 15 years in the operation lifetime is increased by 0.25 RMB/kWh on the basis of the 2005 year on-grid tariff for the newly built coal-fired power plant equipped with desulfurization system. The subsidy of 0.25 RMB/kWh shall be removed for the rest of the operation lifetime. Since 2010, the subsidy price of new approved and established nuclear generated power project would decrease by 2% of which approved and established in last year. The on-grid power price of solar power, oceanic generated power and geothermal power uses governmental-fixed price. The power price standard is prepared by price department of State Council according to reasonable cost plus reasonable profit.

The surcharge on renewable energy power price is calculated as:

Surcharge on renewable energy power price = total amount of renewable energy based power price surcharge/national total amount of power price surcharge

total surcharge on renewable energy based power price $=\Sigma$ ((power price of renewable energy power-provincial desulfurization coal-fired power benchmark on-grid price) × grid purchased renewable power +(operation and maintenance costs of public independent renewable energy power system-average power price of provincial grid × power sales of public renewable energy independent power system) + grid connection fee and other reasonable fee for renewable energy power project)

Of which: 1) national surcharged electricity amount = total electricity sale volume for provincial and above power grid enterprises – power amount for agricultural use – electricity amount sold by Tibet power grid. 2) renewable energy based power amount purchased by power grid = planned renewable energy based power amount – electricity used by power plants. 3) operation and maintenance costs of public independent renewable energy power system = operation cost of public independent renewable energy power system × (1+VAT rate). 4) grid access fee of renewable energy based power generation project and other reasonable costs refers to project investment and operation/maintenance fee specific for power grid access by renewable energy based power generation projects, in accordance with related documents approved by government. Before power transmission and distribution price is approved by the government, the grid access fee will be included into surcharge of renewable energy based power price. Surcharge rate be shared with provincial power grid companies should be identified according to proportion of provincial surcharge power volume to national surcharge power amount, as: surcharge on power price shared with provincial power grid companies = total amount of surcharged renewable energy based electricity × surcharged electricity sale amount within provincial power grid/national surcharged electricity sale amount.

(2) Desulfurization power price policy

To control SO_2 emission of thermal power generator, countermeasures have been issued by the government to promote installation of desulfurization equipments for thermal power generators. In February 1992, the State Environmental Protection Administration announced "Division Program of Acid Rain Control Area and SO_2 Pollution Area", which indicated the important of control of SO_2 emission. It requires under-construction power plants must install desulfurization facilities. It also requires that power plant which sulfur content of coal is higher than 1%, related reduction in SO_2 emission measures should be carried out before 2000, and desulfurization facilities should be installed or related effective emission reduction countermeasures should be conducted before 2010. However, due to lack of incentives or strict constraints, the announcement of this program did not rapidly responded by power producers.

In May 2007, the NDRC and State Environment Protection Department jointly issued "Operation and management approach of desulfurization power price of coal-fired generators and desulfurization equipments" (trial). It pointed out that current coal-fired generator should complete desulfurization transformation according to requirements of "11th Five-year plan of SO_2 control at current coal-fired thermal power plants". After installation of desulfurization facilities, CNY 0.015/kWh would be charged as desulfurization cost recovery rate. The establishment of desulfurization power price policy encouraged power generation companies to install desulfurization facilities. In July 2007, the NDRC issued "Notice regarding to conduct flue gas desulfurization franchise operation pilot work", which proposed desulfurization franchise operation.

(3) Power generation right trading

Power generation right refers to allocated power generation quota obtained in the contract market (annual power generation plan for all generators is approved by government) and mid/short-term trading contract by generation units. Due to some reasons, some units cannot complete their contract power generation volume during the effective period, their power generation quotas can be traded through bilateral/ multi-side transaction or market-matching at an agreed power price. Or units can purchase certain power generation quotas from high-efficient units (hydropower units, large high-efficient thermal power units and etc,.) with agreed power price. For high-efficient units, in addition to complete their own power generation quotas, they can also benefit from power generation quotas trading. To achieve a win-win situation, the marginal power generation cost of transferee must be lower than the marginal cost of transferor. The main cost of marginal power generation is fuel cost. Therefore, the power generation right trading could promote the replacement of inefficient units to eventually reduce total energy consumption and total cost.

(4) Emission right trading

Emission right refers to the right to discharge pollutants within the acceptable quotas allocated by environmental protection agencies, without damage public environment. Large generation units with low coal consumption and high efficiency could reduce pollutant emissions by self-treatment. And the reduced pollutant emission rights could be sold to generators with high coal consumption. For those small generators with high coal consumption and low efficiency, they have to purchase pollutant emission rights to maintain general operation. When their pollutant emission cost is higher the generating revenue, they would have to transfer their power generation rights to large efficient generators.

(5) On-grid power price policy for small thermal power generation units

In January 2007, the state council promulgated the "Options of speeding up decommission of small thermal power generation units", to accelerate decommission of small thermal power generation units. Measures mainly include: to establish large efficient thermal power generation units with high efficiency, low energy consumption and low pollutant emissions; to gradually implement and promote energy-efficient scheduling; to strengthen supervision of grid-connected power price of small thermal power generation units.

In April 2007, the NDRC issued "Notice Regarding to Reduction in On-grid Power Price of Small Thermal Units and Promoting Decommission of Small Thermal Units". It clarified "thermal units with capacity below 50,000 kW, thermal units with capacity below 100,000 kW with 20 years in service, units with capacity below 200,000 kW with expiration of designed service life, and units which on-grid power price is higher than local coal-fired benchmark on-grid power price, should be listed to reduce the power price".

5.2.2 Contribution to energy conservation by power transmission and distribution pricing system

(1) Power transmission replaces coal transportation

During the 11th Five-year plan, 13 large coal bases are planned to be constructed, including Northern Shanxi, Western Inner Mongolia and so on. Most coal bases have the basic condition of large coal-fired power generation. Large thermal power base is far away from power load center. Coal transportation can be replaced by power transmission, to reduce energy consumption during energy transportation. If the power transmission cost has significant advantage than coal transportation cost, the replacement would have positive effects. In this way, the power transmission price system would achieve energy conservation.

(2) Hydropower replaces thermal power at regional level

Replacement between hydropower and thermal power through power transmission price system would promote transfer of power generation efficiency, reduce transmission congestion. There exist great difference between west and east part of China, as well as south and north parts. Besides, the regional load characteristics and energy source structure is various. In this way, grid connection could balance and adjust the power generation situation between west and east parts as well as south and north parts, to reduce reserve capacity, installed capacity and investment in power source construction.

5.2.3 Contribution to energy conservation by sale price

During the peak period, if the generation capacity meets the limitation, peak shaving units with high fuel cost and less utilization hours must be put into operation. Therefore, the production cost would be increased. During the valley period, output power needs to be reduced. Therefore, the high power price happened in the peak time, to reduce power load; and low power price happened in the valley period to encourage reasonable power utilization. Thus, price system should play direct and indirect role of adjustment in all stages, to guide reasonable utilization of power resource by TOU price, reliable price, regional price, differential price and so on.

(1) Differential electricity price

Differential electricity price is mainly applied for energy-intensive enterprises to control their blind development, such as electrolytic aluminum, sodium hydroxide, cement and steel sectors. It aims to promote technological reformation, optimize development of energy-intensive enterprises and save power resource through eliminating preferential electricity price and enhancing electricity tariff for energy-intensive enterprises. In 2004, the NDRC together with State Electricity Regulatory Commission issued differential electricity pricing policy for high energy consumption sectors, including electrolytic aluminum, ferroalloy, calcium carbide, sodium hydroxide, cement and steel. The price will remain the normal level for enterprises to be permitted and encouraged; the price will rise by CNY 0.02/kWh and CNY 0.05 /kWh respectively for enterprises to be eliminated and restricted. In September 2006, to further improve differential electricity price, the State Council extended the application of differential price policy to 8 sectors, included yellow phosphorus and zinc smelting sectors. The standard of differential electricity price will increase gradually in three years. The electricity tariff for enterprises to be eliminated will increase by 50%, to CNY 0.2/kWh from CNY 0.05/kWh. The electricity tariff for enterprises to be restricted will increase to CNY 0.05/kWh from CNY 0.02/kWh. In November 2007, the NDRC issued "Notification Regarding to Further Implementation of Differential Electricity Price Policy". The increased tariff due to differential electricity price policy will be submitted to local treasury which will be included to provincial financial budget, to found a special fund to support local economic structure adjustment and energy conservation work.

(2) Peak/valley TOU electricity price and critical peak price

From the cost point of view, power demand increase by different time will cause different impact on power supply cost. During the peak time, the short-term marginal capacity cost of the system will increase; during the valley time, marginal capacity of the system will not increase. During the peak-load period, power load is close to power generation capacity limit, units for peak shaving (high fuel cost and low utilization hours) must be put into operation, thereby the power generation cost would increase. During the valley-load period, it is needed to decrease power output. Therefore, the electricity tariff at peak period should be higher to encourage peak power generation; and tariff at valley period should be lower to encourage users consumption, to effectively use of power resources and improve load rate of system. If the tariff at peak period is too high, less users might use the electricity during the peak time. If the tariff difference between peak and valley is small, the impact on users would be less and the direction function of pricing system would not be achieved. Currently, the power purchase price is unified calculated. But power sale price is divided by time, which is not conducive to energy saving at power generation side.

On January 1, 2011, the "Power Demand-side Management Approach" was effective. Peak-valley electricity price must be applied according to actual situation of power grid. The division of peak and valley time must be reasonable with principle of "maximum 2 hours of peak period more than valley period". The TOU electricity price is based on average on-grid power price. The peak electricity tariff can be 2-4 times by valley electricity tariff. For critical peak period, critical peak price (CPP) can be applied with appropriately higher price than peak electricity tariff. For the grid side, peak/valley TOU price can be introduced to encourage power plants to fully take advantage of their power generation capacity, on-grid power price at peak period can be appropriate higher, and a little bit lower for on-grid power price at valley period.

(3) High-/Low-flow season power price and seasonable power price

Based on the principle of balance and regulation of power supply in high-flow reason and low-flow reason, on-grid power price and electricity sale price can apply high/low-flow reason power price with appropriate difference between high-flow season and low-flow season. The power price at high-flow season/low-flow season can be lower/higher 30%-50% than average power price. For areas with big difference of seasonal power load, the power price can be adjusted according to different seasons.

(4) High-reliability power price and interchange price

Power supply quality is evaluated by three indicators, including power supply frequency, power voltage and reliability. Frequency quality is assessed by frequency tolerance; voltage quality is measured by voltage flicker, deviation from rating value and distortion of voltage sine wave; reliability of power supply is assessed by time and frequency of power supply by power grid companies. Reliability power price refers to different power price for different users with different requirements in power supply reliability. The power price would be higher if they require higher reliability. Power Law defined that power quality includes power supply reliability. But there is no specific regulation for reliability power price at present.

For users who require high reliability power supply, the power supply companies can adopt multi-power source and multi-loop power supply manner. The users are only responsible for multi-loop fee, there is no difference in power tariff. When the power system requires limitation in power supply, the limit of power supply is sorted by government and nothing to do with power price. The power supply cost is different by different reliability level. To fairly supply electricity, the power tariff must be charged according to actual power supply cost. Reasonable cost for different power use type can be shared by users. Different power tariff should be made for users require different power supply reliability to promote energy conservation.

High-reliability power price should appropriately increase, interchange power price should appropriately decrease. At the peak time, power company can sign agreements with users to reduce or interrupt electricity supply. For those who require high reliability of power supply, power supply would not be interrupted during the peak time by charging high-reliability power tariff to make up capital source of interchange power price.

(5) Ladder-like residential power price

In most places of China, the power price adopts unified electricity tariff no matter how much electricity volume are consumed. The single type power pricing system would result in more compensation is allocated to residents with better economic conditions and big power consumption; while less compensation is given to residents with less electricity consumption. With single type power pricing policy, over electricity consumption would happen due to residential power tariff is low. Implementation of ladder-like residential power price would reduce over power consumption to avoid irrational power consumption, promote energy saving and enhance energy utilization rate.

The power supply cost for power system aiming various users is mainly determined by user's voltage level and power load rate. Higher voltage results in lower power supply cost due to less voltage transformation; otherwise needs higher cost. Power load rate reflects equilibrium of electricity consumption. The power utilization cost is lower for continuous and balanced users. Residential electricity users are mostly end-users, with lowest voltage and concentrated in peak period, therefore, the power supply cost is the highest. In the PRC, the residential power price is mostly achieved through cost-share by increasing of commercial power price. In this way, burden for heavy industries and commercial enterprises would be increased while residents with better financial situation gain more economic compensation. With implementation of ladder-like residential power price, the less electricity consumes the lower power price applies. Thereby, the ladder-like residential power price could reasonable reflect power supply cost, while consider affordability of different users.

(6) Active power price

The "Factor-Adjustment Power Price Approach" is applicable to two-settlement power pricing system and production electricity use for major industrial users. The monthly basic power price and energy power price is calculated by prescribed tariff, then increase or decrease by percentage according power factor-adjustment power price table.

Power factor is weighted average by active power W and reactive power Q consumed in one month:

$$\cos(\varphi) = \frac{W}{\sqrt{W^2 + Q^2}}$$

Based on power factor, every 10 million kW power utilization load, the power grid has to provide users reactive power of 5.8~7.5 million kW to form long distance and big power reactive flow. This is the main reason of big voltage difference and big fluctuation of voltage. When the power factor at user side is low, the transformer capacity would increase and reactive power output from power system would increase.

Transmission power loss:

$$\Delta P = \frac{p^2}{\cos^2 \beta U_n^2} r \times 10^{-3}$$

When the rated voltage Un and output active power P maintains a certain value, supply power loss ΔP is inverse to the square power factor $\cos\beta$, power factor would decrease, power loss would increase. Under the situation of low power factor, to transmit certain active power, the current would increase, the voltage loss would increase. In this way, the power supply quality cannot be ensured. When the current increased, the power supply and distribution equipments cannot be fully utilized, thereby the power consumption would increase. Improvement of power factor of users could reduce line loss, improve voltage quality, and improve utilization rate of electric facilities of users.

(7) Two-settlement power price

According to pricing methods by transaction power or capacity power, power price can be divided into single-part energy price and single-part capacity price. At present, the equipment capacity of users is not considered. The power price is determined only by their power consumption amount which is measured by the meter installed in power distribution room or transformer room. Power grid companies should also provide required capacity for users while meeting requirements of provision of electric energy. Active power loss impacts line loss rate of low-voltage distribution system. But the single-type power pricing system determines the power tariff only by active power amount. And there is no incentives for "low-voltage small industrial" users who don't apply capacity power tariff resulting in the loss of basic power price and power price due to factor adjustment, which is not conducive to users for energy conservation.

Two-settlement power pricing policy divided power price into capacity power price (basic power price) and measured price. Basic power price is based on capacity cost of power plant, which is fixed fee. It is charged as equipment capacity or maximum demand amount multiple basic power price. Measured price is based on energy cost of power plants, which is changeable. It is charged as actual energy consumption multiple related classified power prices. The addition of those two prices is total electricity tariff the user should pay.

5.3 Impacts of EEPGS on on-grid power price

5.3.1 Impacts on on-grid power tariff of decommissioned units and new units

Under the situation of energy-efficient power generation scheduling, the renewable energy ranks at the top of the ranking list. Besides, the hydropower and wind power is stable. Therefore, the utilization of renewable units is relatively stable; the utilization of cogeneration units is constant. But for the thermal units, the utilization hour is decided by decommissioned unit, new unit, and newly added power demand and utilization difference.

1) Impact on decommissioned units

Small generation units have been gradually replaced through LSS program. Their compensated power quotas would be generated by other units (thermal power units). On-grid power price of decommissioned small generation units is normally lower than benchmark power price. After the compensation power capacity policy completed, this amount of power capacity with low price would no longer existed, the grid-connected power price would increase.

2) Impact on newly added generation units

The main beneficiaries of implementation of energy-efficient power generation scheduling policy would be renewable power and heat cogeneration units. This would promote investment to development of renewable resources and speed up decommission of small thermal units and captive power plants. Also, the new scheduling rule will direct the investors to establish large-scale generation units through the LSS program. In this case, the impact on on-grid power price from newly added generating units depends on units' type and their capacity.

(1) **The type of newly added units:** different type of unit cause different impacts on on-grid power tariff. As for renewable energy based units, their investment at early stage is

big; per unit power generation cost is higher than large thermal unit. According to energy-efficient power generation scheduling rule, power generation volume from renewable energy based unit must be fully purchased. Without preferential policy, their on-grid power tariff would increase. As for newly added thermal units, due to less priority is given to them, their on-grid hours are shared with other thermal units. Besides, their on-grid power tariff uses benchmark power tariff. If the average on-grid power tariff of thermal units is higher than benchmark tariff, the on-grid power tariff of newly added thermal units would decrease, otherwise, increase.

(2) **Capacity of newly added units:** the capacity of newly added units can determine whether on-grid power tariff increase or decrease. Normally, larger capacity results in bigger rate of on-grid price fluctuations.

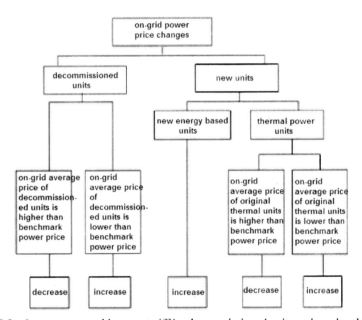

Figure 5-2   Impact on on-grid power tariff by decommissioned units and newly added units

5.3.2   Impacts on on-grid power tariff by energy-efficient scheduling sorting

C_{hi} : capacity of hydropower unit; C_{wj} : capacity of wind power unit; C_{cj} : capacity of thermal power unit; $T_{hi}^{(0)}$: annual on-grid hours of hydropower unit; $T_{wj}^{(0)}$:annual on-grid hours of wind power unit; $T_{cj}^{(0)}$: annual on-grid hours of thermal unit; P_{hi} : annual on-grid tariff of hydropower unit; P_{wj} : annual on-grid tariff of wind power unit; P_{ck} : annual on-grid tariff of thermal unit.

$$\sum_{i \in I} C_{hi}\, T^{(0)}_{hi} + \sum_{j \in J} C_{wj}\, T^{(0)}_{wj} + \sum_{k \in K} C_{ck}\, T^{(0)}_{ck} = Q$$

Where, Q is power generation amount demanded by power grid.

The average on-grid power tariff under energy-efficient power generation scheduling is:

$$P^{(0)} = [\sum_{i \in I} C_{hi}\, T^{(0)}_{hi}\, P_{hi} + \sum_{j \in J} C_{wj}\, T^{(0)}_{wj}\, P_{wj} + \sum_{k \in K} C_{ck}\, T^{(0)}_{ck}\, P_{ck}]/Q$$

After sorted by energy-efficient power generation scheduling rule, the on-grid power generation volume for thermal units can be calculated as:

$$\sum_{k=1}^{s} C_{ck}\, T_{ck} = Q - \sum_{i \in I} C_{hi}\, T_{hi} - \sum_{j \in J} C_{wj}\, T_{wj}$$

Average on-grid power tariff would be:

$$P = [\sum_{i \in I} C_{hi}\, T_{hi}\, P_{hi} + \sum_{j \in J} C_{wj}\, T_{wj}\, P_{wj} + \sum_{k=1}^{s} C_{ck}\, T_{ck}\, P_{ck}]/Q$$

Change of average on-grid power tariff would be:

$$\Delta P = P - P^{(0)}$$

5.4 Impact of EEPGS on cost of power plant and power purchase price

5.4.1 Change of grid investment cost

After implementation of energy-efficient scheduling, the dispatching mode of grid is changed. Once the preferred units are concentrated, the existing power transmission equipments may not meet related load requirements. To meet standards required by hardware, grid company need to increase investment to gird equipments, including construction of new substation, new wires, cables and etc. Since priority is given to renewable energy, in order to achieve large scale development and utilization of wind and solar power, grid connection of renewable energy must be resolved. For Grid Company, great amount of capital need to be invested in grid construction and power transmission, as well as purchasing of wind and solar power. Due to the wind and solar power lacks of stability, large scale connection of wind and solar pwer would impact power structure design, dispatching mode, reactive compensation measure and power quality.

5.4.2　Change of safety management cost

Implementation of energy-efficient scheduling would change the dispatching order of grids. The original network routes and substation might not meet requirements of new dispatching order. Also, more and more renewable energy generation units get involved into the power market, which brought specific security problem to the grid. Take an example of wind power, wind power is not table, large-scale integration of wind power to the grid would cause negative impacts on security operation of power system and relay protection equipments. After implementation of "Energy-efficient power generation scheduling approach", the flow of power grid is decided by unit starting mode, the system operation difficulty would be enhanced; system security would be decreased; investment of grid transformation would be increased to meet security operation requirements under energy-efficient scheduling.

5.4.3　Change of power purchase cost of grid

As shown in Table 5-1, fuel consumption for units with capacity of 100,000 kW is 100g/kWh than units with capacity of 1 million kW. Due to these small units did not install desulfurization equipments, their SO_2 emission rate is 10 times than units with 1 million kW. And their minimum technological output can reach to 80% with poor peak shaving capacity. Normally, the grid price of middle and small coal-fired thermal power generation units is lower. But after introduced environmental friendly units with higher power price standard, the power purchasing cost of Grid Company would be increased. If grid price use national existing approach, after implementation of dispatching, the average grid price would be increased, which could directly impact grid company. Impacts mainly include:

Table 5-1　Typical technical economic index of different types of thermal units
(10,000 kW, g/KWh)

Capacity per unit	Minimum technological output	Fuel per unit	SO_2 emission
100	50%	290	1.6
60	50%	300	2.4
30	60%	320	3.2
20	70%	340	8
10	70%	360	16
5	80%	400	16

(1) Due to the power generated capacity by wind and hydropower units is affected by weather and water amount, power capacity of wind power and hydropower is stable, which would not cause great impacts on Grid Company.

(2) Adjustable hydropower generation capacity would change according to the water condition. If its on-grid price is higher than average on-grid price, the power purchasing cost of grid company would increase; If its on-grid price is lower than average on-grid price, the power purchasing cost of grid company would decrease.

(3) Generation capacity of coal-fired power plant with large capacity and low energy consumption would increase. If its on-grid price is higher than the average on-grid price, the power purchasing cost of grid company would increase; If its on-grid price is lower than the average on-grid price, the power purchasing cost of grid company would decrease.

(4) Generation capacity of coal-fired power plant with small capacity and high energy consumption would increase. If its on-grid price is higher than the average on-grid price, the power purchasing cost of grid company would decrease; If its on-grid price is lower than the average on-grid price, the power purchasing cost of grid company would increase.

(5) For those regular small size coal-fired thermal power plants, according to the system stability and power supply ability, their generation capacity might be decreased. Since the depreciation of fixed cost is basically completed, the power purchasing cost of grid company is lower than average power purchasing cost. Therefore, the power purchasing cost of grid company will increase after reduction of power generation capacity.

(6) Local power source and captive power source mainly includes heat power cogeneration units, residual gas, heat and pressure, coal gangue. Those meet above conditions, the power generation capacity will not be changed a lot; those could not meet above conditions, power generation capacity might be greatly decreased due to impacts by local management policy.

5.4.4 Change of power purchase tariff for power grid

Assume C_i is the installed capacity of unit i; T_i is on-grid hours of unit i; P_i is on-grid tariff of unit i; I_i is power consumption rate of unit i; Q is total power demand; J is integrated line loss rate.

Power purchase average price of power grid companies is $\overline{P}$:

$$\overline{P} = \frac{\sum\limits_{i=1}^{n} C_i T_i P_i (1 - I_i)}{\sum\limits_{i=1}^{n} C_i T_i (1 - I_i)}$$

Of which, $\sum\limits_{i=1}^{n} C_i T_i (1 - I_i) = \dfrac{Q}{1 - J}$.

With consideration of newly added units, decommissioned units and power demand changes:

$$\sum_{i=1}^{n'} C_i' T_i' (1 - I_i) + \sum_{j=1}^{m_1} C_j' T_j' (1 - I_j) = \frac{Q + \Delta Q}{1 - J}$$

Where, C_i' is the installed capacity of unit i with continuous operation; T_i' is on-grid hours of unit i after changes; C_j' is installed capacity of newly added unit j, T_j' on-grid hours of newly added unit j; ΔQ is change value of total power demand.

To maintain power supply-demand balance:

$$\overline{P} = \frac{\sum_{i=1}^{n'} C_i' T_i' P_i' (1 - I_i) + \sum_{j=1}^{m_1} C_j' T_j' P_j' (1 - I_j)}{\sum_{i=1}^{n'} C_i' T_i' (1 - I_i) + \sum_{j=1}^{m_1} C_j' T_j' (1 - I_j)}$$

For decommissioned units, their power generation quotas will be transferred to other units (mainly are thermal units using benchmark power tariff), we assume C_r is installed capacity, T_r is on-grid hours, I_r is power consumption rate, P_r is on-grid power tariff. The impact on power purchase cost from decommissioned units can be calculated as:

$$\Delta F = C_r T_r (1 - I_r)(\overline{P_t} - P_r)$$

$\overline{P_t}$ is average on-grid power tariff of thermal units, when $\Delta F > 0$, indicate power purchase cost increases, otherwise, the purchase case decreases.

Under energy-efficient power generation scheduling, no matter what kind of new units, their grid connection would cause less impacts on on-grid hours of clean energy based units or heat cogeneration units. But they share on-grid hours of thermal units.

For newly added units with capacity of C_n, on-grid hours of T_n, power consumption rate of I_n and on-grid power tariff of P_n, the impact on power purchase cost for power grid companies can be calculated as :

$$\Delta F = C_n T_n (1 - I_n)(P_n - \overline{P_t})$$

$\overline{P_t}$ is average on-grid power tariff of thermal units, when $\Delta F > 0$, the power purchase cost increases, otherwise, it decreases.

6. Energy-Efficient Power Generation Scheduling and Environmental Protection

Energy conservation of power industry in the PRC is promoted by the combination of laws, administrative orders and market incentives, of which the normative documents issued by government at different level play an important role.

6.1 Laws and regulations related to energy conservation

Laws and regulations related to energy conservation in the PRC play a significant role in pollution control, environmental protection and resources saving. Details can be found in Table 6-1.

Table 6-1   Laws related to energy conservation in the PRC

Type	Name	Quantity
Comprehensive	Environmental protection law	1
Pollution control	Air pollution prevention and control law, water pollution prevention and control law, solid waste pollution prevention and control law, environment assessment impact law, ocean environmental protection law, noisy pollution prevention and control law, radioactive pollution prevention and control law	7
Clean production	Clean production promotion law	1
Circular economy	Circular economy law	1
Natural resource and ecological protection	Water law, Sea area utilization management law, agriculture law, fishery law, land management law, water and soil conservation law, mine resource law, forest law, grassland law, power law, renewable energy law, energy conservation law, wild animal protection law, law on prevention and control of desertification	16

(1)   Environmental Protection Law

Environmental protection law is the primary law in environment field. It defines pollution control and control of other hazards, like waste gas, wastewater, solid waste and flue gas. It aims to protect and improve living and ecological environmental and prevent pollution and other hazards through establishment of emission standard and environmental monitoring system, construction of pollution prevention facilities, establishment of release system for excessive discharge.

(2) Electricity Law

General principles of the Electricity Law include: power construction, production, supply and utilization should fully consider environmental protection, by adopting new technology, to reduce harmful substances, prevent pollution and other hazards.

During the construction of power industry, the development planning should be established in accordance with national economic and social development requirements. And the development plan of power industry should be written into national economic and social development plan. The planning should reflect reasonable utilization of energy resource, power source and improvement of economic and environmental benefits. Power transmission and automation communication projects, as well as power grid supporting project and environmental protection projects should be designed, constructed, inspected and put into operation at the same time as power generation projects

(3) Renewable energy law

The state encourages and supports power generation using renewable resources. Special funds for renewable energy development have been established to support scientific and technological research, standards development and demonstration projects of utilization of renewable resources. Preferable tax policy is established for projects listed in development list of renewable nervy resources. The state also provides supporting on the construction of independent renewable energy system in grid disconnected areas. Besides, the state encourages development of biomass fuel, production and utilization of liquid biomass fuel, and installation solar heating system. For the price of renewable energy, the state has the rights to adjust the grid price system of renewable energy based on the principle of promotion on rational utilization and development of renewable energy. And it can be adjusted timely according to the development of renewable energy development technologies.

The amended renewable energy law is effective from April 1, 2010. Based on the original law, the state clarified fully purchasing system of renewable power and establishment of renewable energy development funds.

(4) Energy conservation law

The state encourages and supports development of biogas in rural areas, to promote utilization of biomass, solar and wind resources. Encouragement policies are prepared for energy conservation, including peak-valley electricity pricing, seasonal electricity pricing, interruptible load tariff system, to encourage users of reasonable adjustment of power load. For large power consumption companies, differential pricing system is established with four grades, abandoned, restricted, allowed, encouraged.

(5) Clean production promotion law

In the clean production promotion law, main requirements to power industry include: i) to establish clean production auditing system; ii) to use elimination system for backward production technology, processes, equipment and products which cause serious environmental pollution and waste of resources; iii) to establish scheme of labeling of environmental friendly products, such as energy conservation, water saving, waste recycling, according to related national standards; Iv) to regularly publish the corporate namelist which cause heavy pollution due to over discharge of pollutants.

(6) Circular economy law

The Circular Economy Promotion Law defined that corporate which annual energy consumption capacity and water consumption exceed national standard, such as steel, nonferrous metals, coal, power industry. For those corporate, supervision management system should be established.

6.2   Adjustment of pollutants emission standard for thermal power plants in the PRC

Thermal power industry has great contribution to the national economy. It is also one of major industries which should be responsible for air environmental pollutions, with high emission of SO_2, flue gas and NO_x. Along with increased installed capacity and coal consumption, the environmental pressure caused by power industry would be increased. In this case, whether the pollutants emission from newly added power plants can be controlled, or environmental problems of current power plants can be resolved, is the key point to achieve effective control of air pollution in the PRC. Pollutant emission standard is the technical regulations of environmental protection. It is the technical reference of the implementation of environmental protection law. The strictness of this standard directly impact the power structure, adjustment of power layout, air condition, as well as economic benefits from power industry, development of pollution control technology and industrial development. Establishment and implementation of the emission standard for the national thermal power plants is significantly important to resolve environmental pollution of thermal power industry and healthy development of power industry in the PRC.

Currently, emission standards for thermal power plants in the PRC have been relatively completed. And they have been revised and improved timely according to economic development demand, major contradiction of environmental pollution, and the development of environmental protection technologies. The PRC government has promulgated and implemented four emission standard of air pollutants for thermal power plants, including: "Tentative Emission Standard of Industrial Waste Water, Waste Gas and Solid Waste (GBJ4-73)", "Emission Standard of Air Pollutants for Coal-fired Power Plants (GB13223-91)", "Emission Standard of Air Pollutants for Thermal Power Plants

(GB13223-1996)", and "Emission Standard of Air Pollutants for Thermal Power Plants (GB13223-2003)". Current effective standard is GB13223-2003.

The implementation of Standard GB13223-2003 plays important role in air pollutant emission control, eco-environmental protection and technology promotion in power industry. During the implementation period, the environmental protection program in the PRC has been significantly progressed; the environmental quality of some areas has been improved. However, most Chinese cities still suffer from severe air pollution, with the pattern of complex air pollution. The Chinese government issued a series of new laws, regulations, planning and technical policies, providing higher requirements on environmental protection during the 11th Five-Year Plan. The implementation also promoted the improvement of air pollutant control technology, including thermal power desulfurization, denitration and dust removing techniques. The GB13223-2003 can no longer meet requirements of new challenges. The State Environmental Protection Agency (renamed the Ministry of Environmental Protection in March 2008) issued the revision planning of "Emission Standard of Air Pollutants for Thermal Power Plants" in "Notification Regarding on Annual Revision Plan of Environmental Protection Standards in 2006 " (Office of State Environmental Protection Agency No. 371 (2006)). Currently, the revised version is under comments collection period. The revised version has more stricter limits on SO_2, NO_x and flue gas emission.

(1) The first standard of pollutant emission(GBJ4-73)

The PRC issued the first pollutant emission standard in 1973 – "Tentative Emission Standard of Industrial Waste Water, Waste Gas and Solid Waste (GBJ4-73)", to control SO_2, flue gas emission from chimney. It defined that the allowed emission of each chimney is relevant with geometry height of the chimney, with 3-5 years of compliance deadlines for old power plants. Since this standard allows increasing of emissions limit according to the height of chimney, it stimulus the construction of higher chimney. Even though the high chimney could take advantage of self-purification capacity of the atmosphere, it still cannot solve the pollution problem, or reduce pollutant emission. With the increased capacity of generation units and increased chimney height, the impact area of pollution is expanded, the residence time of pollutants in the air would increase, and also the chance of acid rain is enhanced. In general, this standard has not considered the control of pollutant emission, pollutant discharge density, regional concentration and transfer of pollutants. Furthermore, it has not considered the weather condition, terrain, regional differences and other factors. It failed to effectively control pollutant emissions from power plants.

(2) First version of "Emission Standard of Air Pollutants for Coal-fired Power Plants

(GB13223-91)"

To effectively control air pollutant emission from coal-fired power plants, The State Environmental Protection Agency (renamed the Ministry of Environmental Protection in March 2008) issued " "Emission Standard of Air Pollutants for Coal-fired Power Plants (GB13223-91)" in 1991, effective from August 1, 1992. Improvements of this standard are summarized as below.

In flue gas control: i) discharge density control replaced the discharge capacity control by hour; ii) dust discharge density limits are defined according to new/old power plant and different coal-fire dust; higher ash allows greater emission density; iii) discharge density is classified by filter type (electrostatic precipitator, venture water film precipitator); iv) strict policy is conducted for newly constructed power plants to promote utilization rate of precipitator (average precipitator rate is 98%).

In SO_2 emission control: i) hourly discharge control of the whole plant replaces of hourly discharge control of every single chimney; ii) accepted discharge capacity by chimney's geometric height replaces of by chimney's effective height; iii) it proposed the acceptable SO_2 emission method calculated by equivalent single-source. Which means effective discharge height of multi chimneys equals to effective height of one chimney, to avoid building more chimneys; Iv) it considered the difference of different cities and rural places. The standard is stricter in the urban areas than rural places. This standard plays an important role in reducing acid rain, improvement of environment quality, SO_2 discharge control in industrial pollution, population, pollution source concentrated city areas.

The main problems of this standard include: it doesnot identify long-term indicators; SO_2 emission standard is not strict, without density control. In this standard, only large thermal power plants have the possibility of exceeding the limits. Therefore, it cannot significantly reduce SO_2 emissions.

(3) Amended "Emission Standard of Air Pollutants for Coal-fired Power Plants" (GB1322-1996)

Thermal power plants are the major source of SO_2, and nitrogen oxides, which cause acid rain and SO_2 pollution. The state designated the acid rain and SO_2 pollution control areas to strengthen the control of acid rain and SO_2 pollution. The original "Emission Standard of Air Pollutants for Coal-fired Power Plants (GB1322-91)" has been amended. And the "Emission Standard of Air Pollutants for Coal-fired Power Plants (GB1322-1996)" was issued, effective since January 1997. The characteristics of this standard can be summarized as below:

Firstly, main changes on emission control principles: 1) the control area is enlarged, "thermal power plant" replaced "coal-fired power plant"; 2) according to the construction and operation time of thermal generation units, control plan is prepared for different type of units. Three stages are divided. Considered the implementation time of old standard is not that long, so the two classified power plants are kept, but start another grade after 1997; 3) started strengthening on control of other pollutants which results in acid rain; proposed NO_x emission quality density limits for units ($\geqslant$1000t/h) of power plants belongs to stage 3; 4) newly constructed power plants located in "two control" areas are requires to install fixed flue gas monitoring equipments.

Secondly, continuous strict pollutants control: 1) flue gas: power plant belongs to stage 3, and plant belongs to stage 1 and located in the county and above cities, having 10 years service life since 1 January 1997. The seven-graded method to identify flue gas emission quality density was cancelled. The emission quality control of high-medium coal ash has been strengthened, to promote dust removal equipments installation in newly built plants (average rate is 99%). SO_2: both total emission amount and emission quality control has been established for power plants located in "two control" areas belonging to stage 3. They could only meet the requirements if they use low sulfur coal (sulfur proportion $<$ 1%); 2) the control policy aimed to power plants belongs to stage 3 is more strict in urban areas than rural areas; 3) power plants belong to "two control areas" should also meet requirements of total emission amount control.

Thirdly, the operability of the standard is improved. For old plans with a lot of chimneys, equivalent point source calculation method is adopted to replace ground concentration to calculate SO_2 emission.

However, the new standard still exists some problems: including, P value control standard is not strict enough, which could not significantly reduce SO_2 emissions; proposed SO_2 emission calculation method is complicated, not easy to operate; zoning is not detailed enough; density limits only prepared for SO_2 and NO_x at the stage 3; besides, the density limit for stage 3 is too loose, which cannot play an effective role in the control.

(4) Existing "Emission Standard of Air Pollutants for Thermal Power Plants (GB13223-2003)"

To further improve air quality and control of pollution caused by acid rain, the State Environmental Protection Agency and State General Administration of Quality Supervision, Inspection and Quarantine jointly issued "Emission Standard of Air Pollutants for Thermal Power Plants (GB13223-2003)" in December 2003. Characteristics of the new standard are summarized as below:

— Fully considered the long-term control goal for old and new generation units.

Taking the SO_2 as the example, to determine the emission limit, the new standard considered the goal of control of SO_2 emission for power industry within 5.5 million tons. Therefore, the implementation of emission control is a long term work.

— Emission density limit of SO_2, flue gas, NO_x is proposed and divided into three grades, to integrate the air pollutants control in thermal power plants.

— Clarified the acceptable limit for SO_2 and flue gas in thermal power plants during 2005~2010 by informing in advance, which is conducive for thermal power plants to adopt suitable emission control measures.

— Improved the p value method. Combined the key task of air pollutant emission control during 9th Five-Year Plan and 10th Five-Year Plan, considering unbalanced economic development in the PRC, as well as the geologic difference, the p value method has been improved. It is helpful to adjust the energy source in the PRC and control of environmental impacts from increased power generation amount.

— Acceptable limit mentioned in the new standard is reasonable and scientific. The selection of each control limit is based on mature and reliable technology. And it meets the specific requirements of national west-east power transmission policy and policies of reasonable utilization of resources.

— Considered the time span of stage 1 is big, the SO_2 emission limit in this stage is not applied for each unit, it requires the average emission of units to meet the emission limit. Therefore, the power plants could make flexible adjustment program to meet the requirements of new standard to encourage power plants to select equipments with long service life, large capacity and units with desulfurization equipments, thereby to increase the operability of the standard.

— Fully considered the advanced international experience to establish emission limit which is in line with Chinese characteristics. The emission limit of SO_2, NO_x and flue gas is close to the level required in developed countries, but still far behind with EU and Japan

(5) The "Emission Standard of Air Pollutants for Thermal Power Plants" under comments collection

In the 11th Five-year Plan, the SO_2 emission control has been considered as the key task of pollution control. While in the 12th Five-year plan, the NO_x emission control is proposed to be another key task of pollutants control. Meanwhile, air pollutants emission control in the power industry has been greatly improved. The "Emission Standard of Air Pollutants for Thermal Power Plants"(GB1322-2003) could not meet the requirements of environmental protection any longer. In 2006, the original State Environmental Protection Agency proposed amending plan for this standard. After several years of systematic research, "Emission Standard of Air Pollutants for Thermal Power Plants" (draft for collecting comments) has been prepared to collect comments in 2009 and 2011.

Compared with the currently effective GB1322-2003 standards, the amended "Emission Standard of Air Pollutants for Thermal Power Plants" has significant changes as below. Firstly, the emission limits of SO_2, NO_x and flue gas are greatly decreased, which are now basically at the same level with US, EU, Japan. Therefore, to implement the new standard, advanced emission control technology must be adopted. Secondly, the special air pollutant emission limit is proposed in the new standard. It aims to areas with small environmental capacity and areas where could easily cause serious air pollution. It provides basis to strengthen the enforcement of air pollutants control for power plants in the PRC. Thirdly, mercury emission limit is added into the new standard to control the mercury emission on the power industry.

(6) The change of pollutants emission limits in "Emission Standard of Air Pollutants for Thermal Power Plants"

The amended "Emission Standard of Air Pollutants for Thermal Power Plants" has stricter limits on SO_2, NO_x and flue gas emission. Table 6-2 listed the emission requirements for thermal power plants in the new version of "Emission Standard of Air Pollutants for Thermal Power Plants".

Table 6-2 Emission limits of "Emission standard of air pollutants for thermal power plants" for major coal-fired power plants since 1996(mg/m³)

Operation time of power plants		To July 1992	Aug. 1992 to Dec. 1996	Jan. 1997 to Dec. 2003	Jan. 2004 to Dec. 2011	2012 to now
Flue gas	GB13223-1996	$500^{b,d}$ $1700^{c,d}$	$300^{d,e}$ $1000^{d,f}$		200^g 500^h	
	GB13223-2003	200 ~ 600		50 ~ 500	50	
	New standard ᵃ	30				
SO_2	GB13223-1996ⁱ				1200^j, 2100^k	
	GB13223-2003	1200ˡ		400ˡ	400	
	New standard a	200				100
NO_x	GB13223-1996	Null		650ᵐ		
	GB13223-2003	1100		450 ~ 650		
	New standard a	200		100		

Note: a: in the stage of comment collection; b: emission limits of the boiler using electric dust removal; c: emission limits of the boiler using other dust removal equipments; d: limits for different coal ash; the standard here is for coal ash content between 20% and 25%; e: 670t/h and above, or boiler limits in county or above level; f: 70t/h and below, and boiler limits in county or above level; g: boiler limits in county or above level; h: boiler limits in above county level; i: calculated according to chimney parameter and wind speed, without limits; j: sulfur content is below 1%; k: sulfur content is above 1%; l: before 2010, it is 2100mg/m³; m: 1000t/h and above.

6.3 Energy-efficient scheduling to environmental-protection scheduling

6.3.1 Necessity of extension to energy-efficient environmental-protection power generation scheduling

(1) Power industry will maintain a rapid development trend

Eventhough, the power industry has been greatly developed in recent years, the per capita level still needs further improvement. In 2009, the average installed capacity per capita was only 0.638kW, with an average level of 2659.2kWh per capita, and average residential power usage of 333.6kWh per capita. The per capita installed capacity is about 1/7 of U.S.'s level, 1/4 of Japan's level and 1/3 of Korea's level. The per capita residential power usage is about 1/20 of U.S.'s level and 1/10 of Japan's level. The power supply and demand is basically balanced lacking of stability. In the next 10 years, along with rapid space of industrialization and urbanization, the development of power industry will be speeded up too. According to middle-long term development strategy of energy, it is estimated that generated power in 2020 will reached to 6231.9~7038.2 billion kWh, doubled from current amount, and it will reach to 8086.6~10403.3 billion kWh in 2030. the number will be 8750.6~11947.4 billion kWh in 2050, as shown in Table 6-3.

Table 6-3   Prediction on medium- and long-term installed capacity and power generation quotas in the PRC

	Installed capacity (100 million kW)			Annual average power generation hours (hour)	Power generation amount (100 million kWh)		
	Low-level program	Middle-level program	High-Level program		Low-level program	Middle-level program	High-Level program
2020	15.67	16.7	17.7	3976	62319	66399	70382
2030	20.62	23	26.53	3921	80866	90204	104033
2050	26.51	31.7	36.2	3300	87506	104816	119474

Data source: Research report on medium- and long-term development strategy of power industry in the PRC, Chinese Academy of Engineering

(2) Pollutants emissions in power industry can not be ignored

Power industry has been the target of air pollution control and management. Since the 1970s, the power industry aims flue gas treatment and control as their environmental protection priorities. During the 11th "Five-Year Plan" period, in order to effectively control of SO_2 and acid rain pollution, the PRC for the first time proposed the control target of SO_2 emission amount, and the power industry was aimed as major target. Through engineering emission reduction, structure emission reduction and regulatory emission reduction, the emission of CO_2 in power industry has been greatly decreased; the emission of flue gas has been slightly decreased. However, further efforts should be put in control of nitrogen oxide, as shown in Figure 6-1, the emission of NO_x has been increasing yearly. According to census data of power industry, in 2009, the total emission of SO_2, flue gas and NO_x in the PRC reached to 22.144 million tons, 8.472 million tons and 20.12 tons respectively, and SO_2, flue gas and NO_x from power industry accounted for 45.4%, 39% and 43.9% respectively. In addition, mercury and CO_2 emissions are mainly from coal-fired power plants. The mercury emission of coal-fired power plants accounted for 15% of total mercury emission in the PRC.

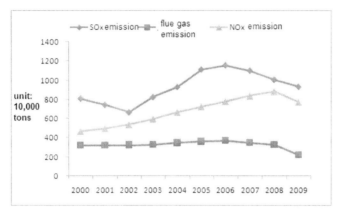

Figure 6-1 SO_2, flue gas and NO_x emissions trends of power industry in the PRC

Generally speaking, reduction of pollutants emissions in power industry relies on two factors: one is decreasing of coal consumption per unit of electricity; the other one is to promote application of pollution treatment equipments and enhance the management operation efficiency. As shown in Figure 6-2, since 1990, consumption of coal equivalent in

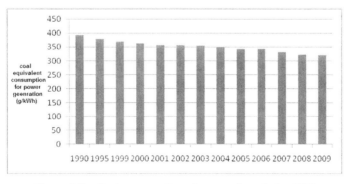

Figure 6-2   Coal consumption of power plants in the PRC

the PRC has been decreased significantly, in 2008, the coal consumption for power generation decreased by 10g/kWh, reaching the largest decline since 1990. Under the condition of increasing in installed capacity of thermal units, and power generation amount, the decrease of SO_2 and flue gas emission has somehow impacted by decrease of coal consumption; by the other hand, if flue gas treatment equipments have not be constructed at a large scale in power plants, the pollution control in power industry would not have significant achievements.

We assume the pollution treatment and control of coal-fired power plants remains the current level, which means to fire one ton coal equivalent will result in SO_2 emission of 22.3 kg, NO_x emission of 7.84 kg, flue gas emission of 7.59 kg, atmospheric mercury emission of 0.000026kg and CO_2 emission of 3640kg. We also assume that in the next 10 years, with encouragement of energy-efficient power generation scheduling and great change of power source structure, the installed capacity of coal-fired units will reach 1 million kW, and the coal equivalent consumption for power generation would reach to 290 g/kWh in 2020. Considerd the above assumptions, in 2020, power industry in the PRC would emit 43 million tons of SO_2, 15.1 million tons of NO_x, 14.8 million tons of glue gas, 50 tons of mercury and 70 billion tons of CO_2, resulting in serious environmental pressure to the PRC.

Therefore, incentives should be taken to reduce coal consumption per unit of electricity and improve pollution treatment of power plants; and energy-efficient power generation scheduling should be further improved, prepare and implement energy-efficient environmental-protection scheduling rule.

(3) Energy-efficient environmental-protection power generation scheduling is the main approach to promote emission reduction in power industry

Since 1970s, the thermal power industry regarded flue gas removal as key assignment of environmental protection. Through 30 years' efforts, the installed capacity of electric dust removal equipments accounted for 95% of total installed capacity, the average dust removal rate can reach above 99% , the flue gas emission in thermal power plants has been effectively controlled. During 11th "Five-Year Plan", large scaled flue gas desulfurization has been carried out in thermal power industry, the proportion of units with flue gas desulfurization equipments increased to 78% in 2009 from 11% in 2005, contributing great efforts in SO_2 control target. The progress of construction of flue gas desulfurization equipments in power industry is shown in Figure 6-3.

With accelerated process of urbanization and industrialization, the rapid growth in NO_x emission resulted in a series of environmental issues, such as, photochemical smog,

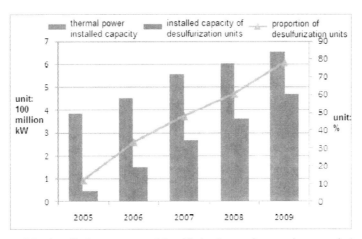

Figure 6-3   Installation progress of desulfurization equipments in power industry

urban fine particle pollution, acid deposition and so on. Thermal power plants are the main source of NO_x, at present, capacity of units with flue gas denitrification facilities is only 50 million kW. With full implementation of NO_x control policy during the 12th "Five-Year Plan", the emission standard is getting stricter, and thermal power industry is regarded as the main sector to conduct NO_x control policy, a large-scale flue gas denitrification construction would be carried out. Currently, aiming to control of particle emissions from thermal power plants, 95% of thermal power plants applied electrostatic precipitation technology, the density of emitted particles maintains at below $50mg/m^3$. Bag-hose precipitation and electric-bag precipitation equipments have a lower installed proportion, only accounting for 5% of total coal-fired installed capacity. As the flue gas emission standard in thermal power industry is getting stricter, dust removal equipments in most power plants cannot meet the new standard ($30mg/ m^3$), certain reformation must be carried out to related equipments. Besides, with increased environmental pressure, in the near future, we will gradually implement emission control on atmospheric mercury, CO_2 and other GHG. Pollution control task for thermal power industry is heavy; the sustainable development of thermal power industry will face great challenges.

Based on above analysis, we may find that pollution control on single aspect cannot meet the requirements of environmental management with increased environmental pressures, collaborative control on SO_2, NO_x, flue gas, mercury and CO_2 should be conducted. From the front-end, mid-end and back-end of pollution treatment point of view, front-end measures, such as decommissioning of small thermal units, speeding up power source structure adjustment is the most effective approaches to achieve energy conservation. However, the pollution treatment in power industry in the PRC still focuses on back-end measures, contributions to energy conservation from structure adjustment and environmental management is relatively small. Energy-efficient environmental-protection power generation scheduling is required to comply with the principle of "energy conservation, environmental protection, economic grounded, and efficiency". Implementation of energy-efficient environmental-protection scheduling would not only promote optimization adjustment of power source structure, promote transformation of "source treatment" to "back-end treatment", but also encourage construction of pollution treatment facilities to effectively achieve energy conservation.

6.3.2 Design of energy-efficient environmental-protection power generation scheduling program

Energy-efficiency power generation scheduling is based on the premise of ensuring reliable electricity supply, in the principle of energy conservation and economic grounded, giving priority to renewable energy. Under this scheduling rule, coal-fired units are sorted according to their integrated energy consumption level. This scheduling rule guarantees the priority of non-fossil energy, while optimizing fossil fuel based units based on equivalent coal consumption, to promote energy conservation and emission reduction in power industry, directly reduce energy consumption level and thereby reduce pollutants emissions.

Energy-efficient environmental-protection power generation scheduling doesnot only consider the energy consumption level, but also environmental factors, such as construction and operation status of pollution treatment facilities, environmental management efficiency. The environmental factors are regarded as constraints. After integration of environmental factors into energy-efficient scheduling, the sorting table of coal-fired units might change somehow. Units with low energy consumption level, installation of pollution control equipments, sound operation management will be given priorities in power generation, to promote development of clean energy based power generation.

(1) Judge reference and standard of energy-efficient environmental-protection power generation scheduling

The content of energy-efficient environmental-protection power generations scheduling is a multi-objective optimization scheduling integrated with energy conservation, environmental protection and economic grounded principles under the premise of reliable power supply. As for the dispatch sorting of different types of generating units, implementation rules should be in line with "Energy-efficient power generation scheduling approach (trial)" issued by the NDRC in August 2007. As for same type of thermal units, especially coal-fired units, sorting table is prepared according to their energy consumption level (coal consumption per unit electricity), giving priority to units with less energy consumption. As for the same type units with same energy consumption level, environmental-protection factors should be considered, as:

Firstly, installation status of pollution treatment facilities; pollution treatment facilities include: precipitation equipment, desulfurization equipment, denitration equipment and mercury removal equipment.

Secondly, environmental performance. Environmental performance refers to pollutants emissions per unit of electricity, pollutants including flue gas, SO_2, NO_x, Hg and CO_2. By implementing the energy-efficient environmental-protection scheduling rule, comprehensive assessment of environmental benefits would be carried out; and power generation priority will give to units with high integrated environmental benefits.

Thirdly, comprehensive benefits. It includes operation status of pollution treatment facilities, environmental cost, economic benefit and operation mechanism of power market. At present, the capacity building of environmental regulatory in the PRC is relatively weak. Power plants pay more attention on construction of pollution treatment facilities, but ignored the operation status of equipments, resulting in un-even flue gas removal and desulfurization efficiency. With gradually improvement of power market mechanism, power generation priority should give to units with high efficiency of pollution treatment and high treatment cost to improve overall operation efficiency of pollution treatment facilities and embody the principles of equity and efficiency.

(2) Sorting method selection of energy-efficient environmental-protection power generation scheduling

With the growth of atmospheric environmental pressure in the PRC, pollution control requirements in power industry are proposed at different stage of environmental protection work. During the 11th "Five-Year Plan", the pollution control aims to flue gas and SO_2, less attention were paid on NO_x control. During the 12th "Five-Year Plan", full implementation of emission control in NO_x will be conducted as well as particle. Therefore, thermal power plants started large scale installation of denitration equipments and improved the efficiency of dust removal equipments. In addition, top five power groups started flue gas mercury removal as well. It can be predicted, after 2015, power industry must carry out CO_2 collection, storage and reuse to reduce CO_2 emissions. Therefore, environmental requirements of power industry is different at different stages, in order to improve pollution treatment level of thermal power plants, energy-efficient environmental-protection power generation scheduling must be implemented. Take the same type of thermal units as example, detailed methods are shown in Table 7-2. It should be noted that the Table 7-2 did not list the energy performance factors as same type of units have the same coal consumption level. Scheduling aimed to different types of units should put energy performance factors into account.

Table 6-4　Sorting of same type of units under energy-efficient environmental-protection power generation scheduling rule

Steps	Condition	Priority sorting	Non-priority sorting
Case 1			
First step	install precipitation+desulfurization+denitration+Hg removal+CCS equipments	yes	No
Second step	compare integrated pollutants emissions performance value	low	High
Third step	compare integrated removal rate of pollutants	high	Low
Case 2			
First step	install precipitation+desulfurization+denitration+Hg removal equipments	yes	No
Second step	compare integrated pollutants emissions performance value	low	High
Third step	compare integrated removal rate of pollutants	high	Low
Case 3			
First step	install precipitation+desulfurization+denitration equipments	yes	No
Second step	compare integrated pollutants emissions performance value	low	High
Third step	compare integrated removal rate of pollutants	high	Low
Case 4			
First step	install precipitation+desulfurization equipments	yes	No
Second step	compare integrated pollutants emissions performance value	low	High

6.3.3　Case studies of energy-efficient environmental-protection power generation scheduling

As for the thermal power plant, factors like coal consumption, desulfurization cost, denitration cost, dust removal cost might impact its power generation scheduling results. As for same type of generating units, the scheduling results rely on environmental treatment cost and pollutants emissions; for the different type of units, power generation scheduling results might be affected by coal consumption, environmental treatment cost and pollutants emissions. This paper designed two programs to analyze environmental economic benefits generated by different types of thermal units, and proposed energy-efficient environmental-protection power generation scheduling program.

(1) Sorting order of same type units by different pollution treatments

Case 1: assume that thermal units with capacity of 1 million kW only installed flue gas precipitation equipment; flue gas removal equipment uses electrostatic precipitators, precipitation rate is 99%.

Case 2: assume that thermal units with capacity of 1 million kW installed flue gas precipitation equipment and wet desulfurization equipment; integrated desulfurization rate is 85%.

Case 3: assume that thermal units with capacity of 1 million kW installed flue gas precipitation equipment, desulfurization and denitration equipment; denitration uses SCR method, denitration rate is 80%.

Table 6-5   Environmental economic benefits of different pollution treatment of same type of generating units

		Case 1 (precipitation)	Case 2 (precipitation +desulfurization)	Case 3 (precipitation +desulfurization +denitration)
Generated electricity (10,000kWh)		500000		
Coal consumption per unit of electricity (g/kWh)		280		
Coal consumption (10,000t)		196		
Pollutants emissions (10,000t)	Flue gas	1.08	1.08	1.08
	SO$_2$	3.14	0.47	0.47
	NO$_X$	0.79	0.79	0.20
	Hg	$0.28*10^{-4}$	$0.14*10^{-4}$	$0.13*10^{-4}$
	CO$_2$	510	510	510
Comprehensive environmental cost (CNY 10,000)	Pollution treatment cost	1195	4068	7989
	Sewage charges	3005	1405	1047
	Economic benefit of gypsum		448	448
	Total	4200	5026	8588

Conclusions can be summarized as blow through estimation of environmental and economic benefits under different pollution treatment measures for thermal units with capacity of 1 million kW:

1) Environmental cost includes pollution treatment cost, pollutants charges and economic benefits generated by selling gypsum generated from desulfurization process. The calculation result shows that the environmental cost of case study 1 is the least, and environmental cost of case study 3 is the highest.

2) As for the pollutants emissions, due to different enterprises adopt different pollution treatment measures, the pollutants emissions of SO$_2$, NO$_x$ and flue gas are significantly different. The total pollutants emission under study case 1 is the highest, and lowest for

study case 3.

3) Under the energy-efficient environmental-protection scheduling, the major factors to determine power generation priority of power plants are coal consumption and environmental benefit. For the same type of generating units, the primary factor to determine the power generation priority is environmental benefit, including pollutant emissions, and environmental cost. Based on above analysis, for the same type of units, priority of power generation should give to units with less pollutants emissions and high environmental investment.

(2) Sorting order of different types of units according to same pollution treatment

At present, power generation scheduling for different types of thermal units only considered the coal consumption difference, without full consideration of difference in environmental economic benefits generated by pollution treatment, which affected the fairness and reasonableness of power generation scheduling.

Assume following three different types of thermal units have the same power generation quotas, pollution treatments include precipitation, desulfurization and denitration, following results can be concluded.

Table 6-6   Environmental economic benefits of different types of units under same pollution treatments

		1 million kW	600,000 kW	300,000 kW
Power generation (10,000 kWh)		500000	500000	500000
Power generation hours (h)		5000	8333	16667
Coal consumption per unit of electricity (g/kWh)		280	324	345
Coal consumption (10,000t)		196	227	241
Pollutants emissions (10,000t)	Flue gas	1.08	1.25	1.33
	SO_2	0.47	0.54	0.58
	NO_x	0.20	0.34	0.45
	Hg	$0.13*10^{-4}$	$0.15*10^{-4}$	$0.15*10^{-4}$
	CO_2	510	590	628
Comprehensive environmental cost (CNY 10,000)	Pollution treatment cost	7989	12528	16373
	Sewage charges	1047	1282	1413
	Economic benefit of gypsum	448	518	552
	Total	8587	13292	17234
Coal cost (CNY 10,000)		12.2	14.2	15.1

Conclusions can be summarized as blow through estimation of environmental and economic benefits under same pollution treatment measures for thermal units with capacity of 1 million kW, 600,000 kW and 300,000 kW:

1) As for the coal cost, under same power generation amount, due to different type of unit has different coal consumption level, resulting in significantly difference between coal consumption and coal cost. The calculation results shows that coal cost of units with capacity of 1 million kW is the lowest, highest for units with capacity of 300,000 kW.

2) As for the environmental cost, comprehensively considered pollution treatment cost, pollutant emission charges and economic benefits generated by selling gypsum generated from desulfurization process, under same power generation amount, the unit with bigger capacity has smaller environmental cost.

3) As for the pollutants emissions, small units consume more coal resulting in higher intensity of pollutant emissions per unit of electricity, thereby results in higher total pollutants emissions. In contrast, large-capacity units consume less coal resulting in lower intensity of pollutant emissions per unit of electricity, thereby results in lower total pollutants emissions

4) Considered both coal consumption and environmental benefit, after implementation of energy-efficient environmental-protection scheduling, under same power generation amount, power generation priority should give to large-capacity units with low energy consumption, low pollutants emissions and low environmental cost.

7. Energy-Efficient Environmental-Protection Scheduling and Its Cost Efficiency Analysis

The coal-dominated energy structure caused great impacts on the natural environment in the PRC. The situation of highly depending on coal for thermal power plants raised the attention from national energy regulatory departments. The government established a series of active countermeasures to enhance energy efficiency and reduce pollutant emission. To evaluate the contribution and potential of those energy conservation policies, environmental performance of thermal power plants should be analyzed. Sensitivity difference among different regulatory policy should be compared to improve related policies.

7.2 Efficiency analysis of energy-efficient power generation scheduling in the PRC

7.2.1 Research method

Data Envelopment Analysis(DEA) is an important method used to evaluate environmental performance. It is characterized as multiple varies capacity, simple and can be easily understood (Yang Hongliang, et, al 2009). DEA analysis based data processing technology is widely applied in verification research of environmental performance, for example, Yang Hongliang and Shi Dan (2008) compared energy efficiency difference of different areas in the PRC by using DEA and nonparametric method-stochastic frontier analysis; Yang Hongliang et al (2009) analyzed impact on energy efficiency from environmental factors; Yang and Pillitt (2009) compared the difference between introduction of undesirable output into DEA and introduction of uncontrollable variables into DEA, and carried out empirical study of thermal power industry in the PRC. Yang and Pillitt (2010) pointed out that strong disposability and weak disposability of undesirable output have significant meanings to efficiency evaluation of thermal power plants in the PRC. However, DEA efficiency analysis technique requires introduction of equation explaining returns to scale as one of constraints. A common practice is to make certain assumptions on returns to scale, and those assumptions are usually a priori. For example, Yaisawarng and Klein (1994) assumed the returns to scale of production technology is constant, variant, non-increasing when they use DEA technique to analyze efficiency of power plants in the USA. Based on the assumption of variant returns to scales (VRS), Golany et al (1994) analyzed power plant efficiency in Israel; Olatubi and Dismukes (2000) carried out performance evaluation of thermal power plants in the USA; Lam and shui (2001) carried out empirical research on efficiency of thermal power plants in the PRC. Based on the assumption constant returns to scale (CRS), Jamasb et al(2004) analyzed impacts on efficiency and performance of power plants in the USA from regulatory policy; Nag (2006) studied carbon emission baseline of power plants in India. Based on VRS and CRS assumption, Jamasb and Pollitt (2003) analyzed impacts on power plants in Europe from power regulatory reform; Pombo and Taborda (2006) analyzed impact difference on power plants' performance, energy efficiency and production efficiency due to power regulatory reform in the USA; Abbott (2006) studied efficiency difference under different power regulatory reform by taking power plants in Austria for example.

Along with increased attention on environmental problems, the Data Envelopment Analysis (DEA) has been greatly applied in environmental performance evaluation from efficiency evaluation of corporate input/output. Related indexes from production are divided into inputs, desirable outputs and undesirable outputs to obtain related efficiency index. Based on research points, related indexes can be divided into two categories: According to the difference of key emphasis, two types of index can be divided: environmental performance index (EPI) is only for environmental evaluation; mixed environmental performance index (MEI) could evaluate both environmental efficiency and economic outputs (Zhou et al., 2008). Efficiency index could provide related information of relevant efficiency, to screen decision making units which need be put more attention on. Xia et al(2011) pointed out among three energy issues, energy security, energy efficiency and energy emission, energy emission and energy efficiency can explain most difference of industries in the PRC.

7.2.2 Module construction

Most pollution problems came from undesirable outputs along with processing of desirable outputs. We can use environmental index to give efficiency evaluation for decision makers. According to the difference of key emphasis, two types of index can be divided: environmental performance index (EPI) is only for environmental evaluation; mixed environmental performance index (MEI) could evaluate both environmental efficiency and economic outputs (Zhou et al., 2008). Efficiency index could provide related information relative to efficiency for decision makers, to screen decision making units which need be put more attention. Xia et al (2011) indicated energy discharge and energy efficiency could explain most difference in the PRC's industry department.

Normally, the returns to scale of environmental DEA technology include constant returns to scale (CRS), non-increasing returns to scale (NIRS) and variant returns to scales (VRS). Based on different assumptions of returns to scale, current environmental performance measure can be divided into two types: one is, independently considering the reduce possibility of undesirable outputs, which means pure environmental performance index (PEI); the other one is: considering integrated efficiency of both desirable outputs and undesirable outputs, which is mixed environmental performance index (MEI).

Environmental DEA technique is based on linear planning technology. Assuming there are $k = 1, 2, ..., K$ decision-making units (DMUs). The input, desirable output and undesirable output of k is $X_k = (x_{1k}, x_{2k}, ..., x_{NK})$, $Y_k = (y_{1k}, y_{2k}, ..., y_{MK})$, $U_k = (u_{1k}, u_{2k}, ..., u_{JK})$, respectively, of which $\sum_{j=1}^{J} u_{jk} > 0$ $(k = 1, 2, ..., K)$, $\sum_{k=1}^{K} u_{jk} > 0$ $(j = 1, 2, ..., J)$. N, M,

J refers to total amount of input, desirable output and undesirable output respectively. Λ is assumption of different returns to scale, $\Lambda = C$ means constant returns to scale (CRS), $\Lambda = NI$ means non-increasing returns to scale (NIRS), $\Lambda = V$ means variant returns to scales (VRS). The MEI can be defined as following linear planning issues under different assumptions:

$$PEI_{\Lambda} = \min \lambda \tag{1}$$

$$s.t. \sum_{k=1}^{K} z_k x_{nk} \leqslant \beta x_{n0}, n = 1, 2, ..., N \tag{2}$$

$$\sum_{k=1}^{K} z_k y_{mk} \geqslant y_{m0}, m = 1, 2, ..., M \tag{3}$$

$$\sum_{k=1}^{K} z_k u_{jk} = \lambda u_{j0}, j = 1, 2, ..., J \tag{4}$$

$$\text{Returns to scale constraints} \tag{5}$$

$$z_k \geqslant 0, k = 1, 2, ..., K \tag{6}$$

Of which, (2) refers to input constraint condition, (3) is desirable output constraint condition, (4) is undesirable output constraint condition and (6) coefficient constraint condition. Constraints under different assumption of returns to scale and β coefficient definition are shown in Table 7-1.

Table 7-1   PEI coefficient definition under different assumption of returns to scale

	β	Constraints of returns to scale
$\Lambda = C$	$\beta = 1$	non-constraint
$\Lambda = NI$	$\beta = 1$	$\sum_{k=1}^{K} z_k \leqslant 1$
$\Lambda = V$	$\beta \leqslant 1$	$\sum_{k=1}^{K} z_k = \beta$

The PEI_Λ bigger value of DMU means better performance in controlling of pollutants and better environmental performance. PEI_C, PEI_{NI} and PEI_V can also be used to judge the return of scale to DMU. If $PEI_C = PEI_V$, CRS happens. If $PEI_C \neq PEI_V$, then PEI_{NI} would be further studied. If $PEI_{NI} = PEI_V$, decreasing of returns to scale happens.

Under VRS situation, based on DEA technique, MEI can be defined as:

$$MEI = \min \frac{\lambda}{\theta} \tag{7}$$

$$s.t. \sum_{k=1}^{K} z_k x_{nk} \leqslant x_{n0}, n = 1,2,...,N \tag{8}$$

$$\sum_{k=1}^{K} z_k y_{mk} \geqslant \theta y_{m0}, m = 1,2,...,M \tag{9}$$

$$\sum_{k=1}^{K} z_k u_{jk} = \lambda u_{j0}, j = 1,2,...,J \tag{10}$$

$$\sum_{k=1}^{K} z_k = 1 \tag{11}$$

$$\theta, z_k \geqslant 0, k = 1,2,...,K \tag{12}$$

The calculation of MEI is similar with that of conventional input/output. But the conventional input/output calculation compares the efficiency between input and output, while calculation of MEI compares the efficiency between desirable output and undesirable output. If MEI = 1, then DMU_0 is environmental efficient; if MEI<1, DMU_0 is environmental inefficient. Zhou et al.(2008) proved:

$$MEI = \min \lambda \tag{13}$$

$$s.t. \sum_{k=1}^{K} z_k x_{nk} \leqslant \beta x_{n0}, n = 1, 2, ..., N \qquad (14)$$

$$\sum_{k=1}^{K} z_k y_{mk} \geqslant y_{m0}, m = 1, 2, ..., M \qquad (15)$$

$$\sum_{k=1}^{K} z_k u_{jk} = \lambda u_{j0}, j = 1, 2, ..., J \qquad (16)$$

$$\sum_{k=1}^{K} z_k = \beta \qquad (17)$$

$$z_k \geqslant 0, k = 1, 2, ..., K \qquad (18)$$

Noted that $\lambda = \beta = 1$ is the solution of above equation, therefore, the value of MEI ranges between $(0,1]$. Bigger value of MEI means better environmental efficiency.

7.2.3 Data source and primary analysis

In March 2009, the State Electricity Regulatory Commission together with China Electricity Council published statistic data of large power plants in 2008. Related data can be downloaded from official website of State Electricity Regulatory Commission. Data used in the paper also is sourced from there. By the end of 2008, in total 30 power plants which installed capacity has exceeded 2 million kW, total installed capacity reached to 557.02 million kW accounting for 70.28% of total installed capacity in the PRC, of which, installed capacity for thermal power units and clean energy based units (natural gas and biomass) reached to 455.48 million kW and 115.23 million kW respectively.

By the end of 2008, among all 30 power plants with installed capacity over 2 million kW, 12 plants are state-owned, and 18 power plants are invested by local or foreign capitals. Units with capacity of 600,000 kW and above reached to 329, with total capacity of 211.4 million kW. Installed capacity of 12 state-owned power plants reached to 442.27 million kW, accounting for 55.80% and 79.40% of total installed capacity in the PRC and that of 30 power plants. The total installed capacity of five top power plants in the PRC accounted for 44.57%, 63.42% and 79.88% of that of China, 30 power plants and 12 state-owned plants respectively. The total installed capacity of 18 power plants with diversified investment reached to 114.75 million kW, accounting for 14.48% and 20.60% of that of China and 30 power plants. The national power generation capacity is highly concentrated. The power generation production capacity of state-owned power plants accounted for 55.81% of total capacity in the PRC; the thermal power capacity accounted for 58.04% of total thermal power production capacity in the PRC; the thermal power generation quantity accounted for 57.43% of total thermal power generation amount in the PRC. The concentration of thermal power generation production capacity determined the necessity of environmental performance evaluation and environmental regulatory.

As for the data source, regarding to pollutant emission data of power plants (flue gas, nitrogen oxides data of China Huaneng Group and China Resources Power Holdings Company Limited is missing; pollutant emission data of Henan Investment Group is missing), in 2008, the largest absolute volume of SO_2 and dust emissions happened in China Guodian Corporation, reaching to 1.285 million tons and 222,000 tons respectively. The largest emission volume of nitrogen oxide was from Datang Group, reaching to 800,000 tons. Figure 1 illustrates the density of pollutant emissions of major thermal power plants in 2008. From the pollutant emission density of view, companies which SO_2 density exceeded 200 tons/100 million kWh include: China Guodian Corporation (462.16/100 million kWh), China Huadian Corporation (450.71 tons/100 million kWh), China Power Investment Corporation (391.80 tons/100 million kWh), China Huaneng Group (325.16 tons/100 million kWh), Sichuan Investment Group (300.68 tons/100 million kWh), Guangxi Investment Group (285.51 tons/100 million kWh), Anhui Energy Group (226.19 tons/100 million kWh), Hebei Construction Investment Corporation (206.87 tons/100 million kWh) and Zhejiang Energy Corporation (204.56 tons/100 million kWh). Companies which flue gas emission density exceeded 50 tons/100 million kWh include: China Guodian Corporation, China Huadian Corporation, China Power Investment Corporation and Anhui Energy Group. The highest density happened in Anhui Energy Group. Companies which nitrogen oxides exceeded 100 tons/100 million kWh include: Datang Group, China Guodian Corporation, China Huadian Corporation, China Power Investment Corporation, Zhejiang Energy Corporation, Shenhua Group, Shandong Luneng Group, Guizhou Jinyuan Group, Hebei Construction Investment Corporation, Jiangsu Guoxin Group, Shenneng Group, Shenzhen Energy Corporation, Hubei Energy Corporation, Sichuan Energy Corporation, Anhui Energy Corporation and Guangdong Development Group. The highest density of nitrogen oxides happened in Jiangsu Guoxin Group. The density of pollutants emissions for power plants with large installed capacity is normally higher than other power plants. Also, coal quality and geographical difference may result in difference in pollutant emissions. The extension of enterprises would not necessarily lead to reduction in density of discharged pollutants; and simple corporate restructuring cannot improve environmental regulatory efficiency of thermal power plants and eventually improve energy efficiency in thermal power plants in the PRC neither.

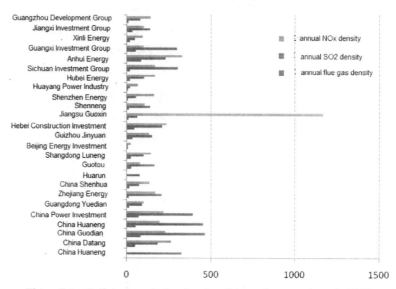

Figure 7-1   Pollutants emission density of thermal power plants in 2008

Note: flue gas, nitrogen oxides data of China Huaneng Group and China Resources Power Holdings Company Limited
is missing; pollutant emission data of Henan Investment Group is missing

7.2.4   Efficiency measurement and analysis

(1) Distinguish of efficiency measurement situation

Based on data envelopment analysis technology, economic and environmental efficiency of major power plants in the PRC has been evaluated. We considered coal consumption, SO_2 emissions, flue gas emissions, nitrogen oxide emissions as undesirable outputs, to obtain related environmental performance index (EPI). Also, empirical research of returns to scale for power plants was carried out. Meanwhile, based on mixed environmental performance index (MEI), comparison analysis of relative efficiency of thermal power plants in the PRC was conducted.

Due to lack of flue gas and nitrogen and oxide emission data of China Huaneng Group and China Resources Power Holdings Company Limited, data sources are divided into two kinds of situation to discuss energy utilization efficiency and environmental benefit. We calculated the efficiency of power generation companies using different indicators. Matlab software is adopted. During the actual analysis, we consider the installation capacity as the replacement of capital, and input in DEA[12]; generating capacity as desirable outputs,

---

12   In the actual calculation, we didn't include laboring into efficiency evaluation, due to the data is hard to obtain;also
during the current employment situation in the PRC, the laboring would not be considered as a constraint.

undesirable outputs include following four situations: coal consumption[13]; coal consumption, SO_2; coal consumption, SO_2 and flue gas; coal consumption, SO_2, flue gas and NO_x. Of which, the coal consumption = coal consumption per unit of electricity * power generating capacity. Detailed information can be found in Table 7-2.

Table 7-2   Undesirable output and data difference under different cases

	Sample difference	Undesirable outputs
Case 1	Include Huaneng and Huarun	Coal consumption
Case 2	Include Huaneng and Huarun	Coal consumption, SO_2
Case 3	Exclude huaneng and huarun	Coal consumption
Case 4	Exclude huaneng and huarun	Coal consumption, SO_2
Case 5	Exclude huaneng and huarun	Coal consumption, SO_2 and flue gas
Case 6	Exclude huaneng and huarun	Coal, SO_2, flue gas and NO_x

(2) Returns to scale

By using above data processing method, we can obtain PEI and MEI under different returns to scale assumption. The Table 7-2 described the returns to scale. It can be found that: if only consider the coal consumption as undesirable output, which means aiming to the goal of energy conservation, most companies has a decreased retunes to scale. Those companies has an increased returns to scale are small companies. If consider coal consumption, SO_2, flue gas and NO_x as undesirable outputs, which means aiming to both economic benefits and environmental benefits, the returns to scale for most companies remain the same value. For the first case, 13 companies have decreased returns to scale value. For the second case, 17 companies remain the same. Under both situations, only Shenneng Group remains the same. Under both situations, only Huadian Group reflects decreased returns to scale value, while Hubei Energy Group reflects an increased returns to scale value.

The increased returns to scale are mainly resulted from intensive and effective use of resources; decreased returns to scale are mainly resulted from reasonable coal source and management efficiency. The government should control the power plants' scale through administrative measures or industrial policies. For companies with increased returns to scale value, the government should expand its scale. For companies with decreased returns to scale value, the government should encourage them to improve efficiency and reduce its scale. Therefore, the government must reasonable select the regulatory content to optimize

---

13   Consideration of coal consumption as undesirable outputs is inspired by Lozano and Gutiérrez (2008).

the scale of power plants.

Table 7-3   Returns to scale value under different situations

Company	Case 1	Case 2	Case 3	Case 4	Case 5	Case 6
C1	D	D				
C2	D	D	D	D	C	D
C3	D	C	D	C	C	C
C4	D	D	D	D	D	C
C5	D	D	D	D	D	C
C6	D	D	D	D	D	D
C7	D	D	D	D	C	C
C8	D	D	D	D	D	D
C9	D	D				
C10	D	D	D	D	D	C
C11	C	C	C	C	D	C
C12	C	C	C	C	D	C
C13	D	D	D	D	D	D
C14	D	C	D	C	D	D
C15	C	D	C	D	D	C
C16	C	C	C	C	C	C
C17	D	C	D	C	C	C
C18	I	C	I	C	C	C
C19	I	I	I	I	I	I
C20	I	I	I	I	C	C
C21	I	I	I	I	C	C
C22	I	I	I	I	I	C
C23	I	I	I	I	C	C
C24	I	I	I	I	I	C
C25	I	I	I	I	C	C

Note: C means returns to scale does not change; I means returns to scale increased; D means returns to scale decreased.

(3) Efficiency ranking and stability

A ranking list of companies is sorted according to MEI value. The table 7-4 shows the efficiency ranking result. It can be found that the sorting is sensitive of evaluation content. Such as, if it only considers the coal consumption, the China Power Investment follows to the bottom of the list. If it considers Coal consumption, SO_2 and flue gas, it ranks towards the top. The similar situation happened to Hebei Construction Investment Company too.

Since the ranking position has great impacts on company leader and future development. The environmental efficiency has become an important evaluation indicator.

Therefore, a reasonable evaluation indicator system is very important to evaluate the company efficiency, to encourage them improve economic and environmental efficiency. During the establishment of evaluation indicator system, the government should comprehensively consider regional, economic, environmental and social factors to achieve maximum economic, environmental and social benefits.

Table 7-4 Efficiency ranking list under different situations

Company	Case 1	Case 2	Case 3	Case 4	Case 5	Case 6
C1	15	9				
C2	4	17	14	17	1	22
C3	20	1	18	1	1	1
C4	7	7	4	7	13	1
C5	23	12	21	11	1	1
C6	13	18	12	16	16	1
C7	10	11	9	13	21	1
C8	9	14	8	10	20	1
C9	17	20				
C10	18	22	16	20	1	23
C11	22	23	20	21	22	1
C12	8	1	7	1	15	1
C13	24	25	22	23	14	1
C14	21	21	19	19	1	1
C15	11	15	10	14	1	15
C16	1	1	1	1	12	20
C17	3	1	3	1	1	1
C18	5	1	5	1	18	1
C19	16	1	15	12	1	18
C20	6	10	6	8	1	17
C21	12	13	11	15	23	21
C22	14	24	13	18	17	1
C23	25	8	23	22	19	1
C24	19	19	17	9	1	19
C25	1	16	1	1	1	16

Note: companies at the top of the list have higher efficiencies.

7.2.5 Conclusion

By using Data Envelopment Analysis method, the environmental performance of power plants in the PRC has been analyzed through comparison of environmental performance measurements under different returns to scale assumptions. The returns to scale of power plant have been obtained. And a ranking list of power plants' performance has been prepared according to their environmental performance measurements. Through this study, it can be found that company scale doesnot affect the efficiency directly. Some medium-scale company has higher efficiency. While the top five power generation groups have no significant advantages in terms of efficiency. The returns to scale for the company rely on the evaluation of outputs. The efficiency ranking of the company is sensitive to the evaluation content. Therefore, a reasonable evaluation indicator system is very important to evaluate the company efficiency, to encourage the company improve economic and environmental efficiency. Due to some data is hard to obtain, this paper did some approximate processing of the data. If all data can be obtained, more significant conclusion might be found.

Appendix: samples and its code

Code	Company	Code	Company name
C1	China Huaneng	C14	Hebei Construction and Investment Corporation
C2	China Datang	C15	Jiangsu Guoxin
C3	China Guodian	C16	Shenneng Group
C4	China Huadian	C17	Shenzhen Energy Group
C5	China Power Investment Group	C18	Huayang Power
C6	Guangdong Yuedian	C19	Hubei Energy group
C7	Zhejiang Energy group	C20	Sichuan Investment Group
C8	China Shenhua	C21	Anhui Energy
C9	Huarun Power	C22	Guangxi Investment Group
C10	Guotou Power plant	C23	Xinli Energy Development
C11	Shandong Luneng Development Group	C24	Jiangxi Investment Group
C12	Beijing Energy Investment	C25	Guangzhou Development Group
C13	Guizhou Jinyuan group		

7.3 Regulatory cost evaluation of thermal power plants in the PRC

7.3.1 Research significance

The energy structure of the PRC with coal as the dominant determined the production structure of power plants. With increasing attention on energy and environmental problems, the administrative and economic regulatory plays an important role in promotion of power plants' efficiency and optimization of production structure. Selection of regulatory content is the most important step, which eventually determines the economic and social cost of the regulatory. And it would also impact the regulatory effect.

During the 11th Five-Year Plan, the PRC considered the energy consumption per unit GDP as constraint for national economic and social development; during the 12th Five-Year Plan, in addition to energy consumption per unit GDP, the constraint of economic and social development also includes chemical oxygen demand, SO_2 emission, emissions of ammonia and nitrogen oxide, CO_2 emission per unit GDP. The change of those constraint indicators would impact the production and operation activities of enterprises. Therefore, distinguishing the energy and environmental regulatory cost of thermal power plants in the PRC and exploring the implementation effect of different regulatory policies by different constraint conditions have significant meanings for enhancement of scientific knowledge of policies and objective evaluation.

Empirical evaluation of environmental regulatory cost can be traced back to Färe et al.(1989). Färe et al.(1989) extended the traditional efficiency analysis method (Farrell,1957) by considering regulatory cost of undesirable output as economic performance difference under different assumptions of economic development. Based on the profit maximization assumption, Brännlund et al.(1995) evaluated environmental regulatory cost through profit difference of enterprises under comparison of regulatory condition and non-regulatory condition. Based on linear planning method, Pasurka (2001) analyzed relationship between pollutant regulatory cost and technical changes for manufacturing and power industries in the USA, and found out that proportion of regulatory cost of pollutants which cannot be measured increased due to changes in technologies. Based on directional distance function, Boyd et al.(2002) analyzed impacts on production efficiency and environmental performance from technology changes, and found out regulatory shadow price of nitrogen oxides from glass container production sectors is higher than other sectors. Vardanyan and Noh (2006) studied possible impacts on cost decreasing from different technical function of power industry in the USA, and compared the difference of different cost estimation methods when they were applied in power industry in the USA.

Domestic literatures emphasize on economic impact under environmental constraint. For example, Xu Shichun and He Zhengxia (2007) studied impacts on product quality and profit from environmental regulatory; Wang Bing (2008) carried out empirical study on environmental regulatory and total factor productivity growth of APEC countries; Bai Xuejie and Song Ying (2009) analyzed relation between environmental regulatory and efficiency of thermal power plants in the PRC by using DEA module; Jin Pei (2009) analyzed the relationship between resource environmental regulatory and industrial competitiveness. This paper carried out empirical research on environmental and energy regulatory cost for Chinese thermal power plants for the first time, compared difference of regulatory cost under different constraint condition and found out the regulatory cost distribution under different constraint conditions.

7.3.2 Description of environmental regulatory cost evaluation method

Considered a production process in which desirable outputs and undesirable outputs are jointly produced, assume input vector $x \in R_+^N$ would generate a series of desirable outputs $y \in R_+^M$, along with undesirable outputs $b \in R_+^H$. Assume $k = 1, 2, ..., K$ enterprises use $n = 1, 2, ..., N$ different types of inputs, to obtain $m = 1, 2, ..., M$ desirable outputs, meanwhile generating $h = 1, 2, ..., H$ undesirable outputs. Figure 7-2 shows the relations among actual production level, production frontier under regulatory and production frontier without regulatory.

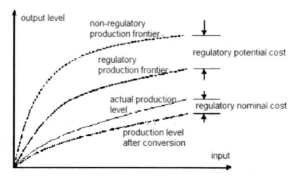

Figure 7-2   Production frontier and regulatory cost

We define the directional distance function as:

$$\beta_w^* = \max \beta_w$$

$$s.t. \sum_{k=1}^{K} z_k x_n^k \leqslant x_n^{k'}, n = 1, 2, ..., N$$

$$\sum_{k=1}^{K} z_k y_m^k \geqslant (1 + \beta_u) y_m^{k'}, m = 1, 2, ..., M$$

$$\sum_{k=1}^{K} z_k b_h^k \geqslant b_h^{k'}, h = 1, 2, ..., H$$

$$z_k \geqslant 0, k = 1, 2, ..., K$$

If the government undertakes regulatory on H_1 undesirable outputs, the desirable outputs of enterprises will decrease, in this case, the directional distance function should be defined as:

$$\beta_s^* \big|_{H_1} = \max \beta_s \big|_{H_1}$$

$$s.t. \sum_{k=1}^{K} z_k x_n^k \leqslant x_n^{k'}, n = 1, 2, ..., N$$

$$\sum_{k=1}^{K} z_k y_m^k \geqslant (1 + \beta_s) y_m^{k'}, m = 1, 2, ..., M$$

$$\sum_{k=1}^{K} z_k b_h^k = b_h^{k'}, h = 1, 2, ..., H_1$$

$$\sum_{k=1}^{K} z_k b_h^k \geqslant b_h^{k'}, h = 1, 2, ..., H - H_1$$

$$z_k \geqslant 0, k = 1, 2, ..., K$$

If the government undertakes regulatory on all undesirable outputs, the directional distance function should be defined as:

$$\beta_s^* = \max \beta_s$$

$$s.t. \sum_{k=1}^{K} z_k x_n^k \leqslant x_n^{k'}, n = 1, 2, ..., N$$

$$\sum_{k=1}^{K} z_k y_m^k \geqslant (1 + \beta_s) y_m^{k'}, m = 1, 2, ..., M$$

$$\sum_{k=1}^{K} z_k b_h^k = b_h^{k'}, h = 1, 2, ..., H$$

$$z_k \geqslant 0, k = 1, 2, ..., K$$

We evaluate the output difference under different regulatory condition based on different directional distance function. The output difference actually reflects the environmental regulatory cost of enterprises. In some cases that regulatory happened to some undesirable outputs, the regulatory impact index of number k enterprise can be defined as:

$$\rho_k = \beta_w^* |_{H_1} - \beta_s^*, \quad k = 1, 2, ..., K$$

Followed, regulatory happened to all undesirable outputs, the regulatory impact index of number k enterprise can be defined as:

$$\rho_k = \beta_w^* - \beta_s^*, \quad k = 1, 2, ..., K$$

7.3.3 Data source and description

In March 2009, the State Electricity Regulatory Commission together with China Electricity Council published statistic data of large power plants in 2008. Related data can be downloaded from official website of State Electricity Regulatory Commission. Data used in the paper also is sourced from there.

Due to lack of flue gas and nitrogen and oxide emission data of China Huaneng Group and China Resources Power Holdings Company Limited, data sources are divided into two kinds of situation to discuss energy utilization efficiency and environmental benefit. We calculated the efficiency of power generation companies using different indicators. Matlab software is adopted. During the actual analysis, we consider the installation capacity as the replacement of capital, and input in DEA[14]; generating capacity as desirable outputs, undesirable outputs include following four situations: coal consumption[15]; coal consumption, SO_2; coal consumption, SO_2 and flue gas; coal consumption, SO_2, flue gas and NO_x. Of which, the coal consumption = coal consumption per unit of electricity * power generating capacity. Detailed information can be found in Table 7-5.

Table 7-5   Undesirable outputs and data difference under different situations

	Sample difference	regulatory content
Case 1	Include Huaneng and Huarun	Coal
Case 2	Include Huaneng and Huarun	Coal, SO_2
Case 3	Exclude huaneng and huarun	Coal
Case 4	Exclude huaneng and huarun	Coal, SO_2
Case 5	Exclude huaneng and huarun	Coal, SO_2, flue gas and nitrogen oxides

---

14 In the actual calculation, we didn't include laboring into efficiency evaluation, due to the data is hard to obtain;also during the current employment situation in the PRC, the laboring would not be considered as a constraint.
15 Considerition of coal consumption as undesirable outputs is inspired by Lozano and Gutiérrez (2008).

7.3.4 Calculation and analysis of regulatory cost

(1) Analysis on regulatory impact index under different situations

Impact index is shown in Table 7-6. We can easily find out: Sichuan Investment Group is heavily impacted by the regulatory, followed by Beijing Energy Investment group. The regulatory content for case 1 is less. The impact under case 1 is relatively small. Among all cases, companies with least loss of production potential accounted for 40~50% of 2008 level include: China Datang, Xinli Energy Development, Jiangxi Investment group. Companies with least loss of production potential accounted for 30~40% of 2008 level include: Guangdong Yuedian, Zhejiang Energy group, China Shenhua, Guizhou Jinyuan group, Jiangsu Guoxin, Shenneng Group, Huayang Power, Hubei Energy group. The rest companies are below 30% of production level in 2008. With different regulatory content, the impact is different.

Table 7-6   Impact index of regulatory under different situations

Company	Case 1	Case 2	Case 3	Case 4	Case 5
C1	0.2638	0.2638	-	-	-
C2	0.4312	0.4312	0.4484	0.4492	0.4492
C3	0	0	0	0	0
C4	0.0731	0.0731	0.0731	0.0731	0.0769
C5	0	0	0	0	0
C6	0.3352	0.3551	0.3354	0.3606	0.3713
C7	0.3658	0.3658	0.3693	0.3693	0.4199
C8	0.3199	0.3441	0.3219	0.3526	0.3526
C9	0.2842	0.3062	-	-	-
C10	0.5221	0.5221	0.5371	0.5374	0.6209
C11	0	0	0	0	0
C12	1.0231	1.0666	1.0231	1.0666	1.0666
C13	0.351	0.354	0.3715	0.3857	0.4398
C14	0.2115	0.2115	0.2034	0.2034	0.2034
C15	0.3988	0.4218	0	0	0
C16	0.3136	0.3136	0.3136	0.3136	0.3136
C17	0.5502	0.5576	0.5545	0.5576	0.5576
C18	0.3877	0.4208	0.3877	0.4208	0.4208
C19	0.3191	0.3359	0.3246	0.3461	0.3461
C20	2.2681	2.2681	2.2681	2.2681	2.3437
C21	0.0986	0.0986	0	0	0
C22	0.2692	0.2692	0.2834	0.2834	0.4179
C23	0.4757	0.4936	0.4821	0.5228	0.5228
C24	0.4747	0.476	0.5034	0.5114	0.5795
C25	0.2744	0.2948	0.2759	0.2968	0.3481

(2) Regulatory output cost in power generation capacity

Table 7-7 shows the cost under different regulatory situations, calculating with power generation capacity. Calculated with power generation capacity, China Datang Group has the biggest cost of coal consumption regulatory. With regulatory content of coal consumption and SO_2, the China Huaneng Group has the biggest cost. Companies with magnitude of regulatory cost above 10 billion degrees include: China Huadian, Guangdong Yuedian, Zhejiang Energy, China Shenhua, Huarun Power, Guotou Power, Beijing Energy, Guizhou Jinyuan, Shenzhen Energy, Sichuan Investment. The regulatory output cost of power generation companies under case 1-5 is 23.87%, 24.32%, 22.37%, 23.37% and 24.28% respectively. The sample quantity of case 1 and 2 is bigger than case 3 and 4, thus the regulatory cost is higher.

Table 7-7 Output cost under different regulatory constraints in power generation capacity
(100 million kWh)

Company	Case 1	Case 2	Case 3	Case 4	Case 5
C1	931.8577	931.8577	-	-	-
C2	1320.49	1320.49	1373.163	1375.613	1375.613
C3	0	0	0	0	0
C4	190.0863	190.0863	190.0863	190.0863	199.9677
C5	0	0	0	0	0
C6	321.7384	340.8392	321.9303	346.1183	356.3886
C7	315.1989	315.1989	318.2147	318.2147	361.8152
C8	307.7557	331.037	309.6798	339.2143	339.2143
C9	258.1673	278.1521	-	-	-
C10	215.5717	215.5717	221.7651	221.889	256.3656
C11	0	0	0	0	0
C12	235.6486	245.6678	235.6486	245.6678	245.6678
C13	108.4721	109.3992	114.8074	119.1957	135.9146
C14	66.5992	66.5992	64.0486	64.0486	64.0486
C15	98.3174	103.9877	0	0	0
C16	85.6128	85.6128	85.6128	85.6128	85.6128
C17	126.6335	128.3367	127.6232	128.3367	128.3367
C18	82.6677	89.7255	82.6677	89.7255	89.7255
C19	9.5878	10.0926	9.753	10.399	10.399
C20	120.3681	120.3681	120.3681	120.3681	124.3802
C21	16.5648	16.5648	0	0	0
C22	29.9027	29.9027	31.4801	31.4801	46.4203
C23	60.8385	63.1278	61.657	66.8623	66.8623
C24	56.3665	56.5208	59.7743	60.7243	68.8105
C25	38.3932	41.2475	38.6031	41.5274	48.7051
Sum	4996.839	5090.387	3766.883	3855.084	4004.248
%	23.8684	24.3152	22.8377	23.3724	24.2768

(3) Regulatory cost in monetary

Table 7-8 shows the regulatory cost in monetary under different constraints. The regulatory cost in monetary considered the grid price difference of different power plants. Calculated with the nominal output, China Datang has the biggest regulatory cost. Companies with regulatory cost over CNY10 billion include: China Huaneng, Guangdong Yuedian, Zhejiang Energy, China Shenhua, and Huarun Power. The grid-connected electricity price of power Generation Company is mainly impacted by the geological location. In 2008, Guangzhou Development Group has the highest average grid-connected power price. The environment and efficiency factors have less impact on power price system.

Table 7-8   Regulatory cost in monetary (CNY 100 million)

Company	Case 1	Case 2	Case 3	Case 4	Case 5
C1	329.1414	329.1414	-	-	-
C2	483.8541	483.8541	503.1544	504.0521	504.0521
C3	0	0	0	0	0
C4	66.8325	66.8325	66.8325	66.8325	70.3066
C5	0	0	0	0	0
C6	148.1187	156.9121	148.2071	159.3425	164.0706
C7	147.7337	147.7337	149.1472	149.1472	169.5828
C8	114.8834	123.5742	115.6017	126.6267	126.6267
C9	105.3529	113.5083	-	-	-
C10	68.1207	68.1207	70.0778	70.1169	81.0115
C11	0	0	0	0	0
C12	80.7048	84.1362	80.7048	84.1362	84.1362
C13	-	-	-	-	-
C14	23.4429	23.4429	22.5451	22.5451	22.5451
C15	39.8884	42.1888	0	0	0
C16	35.5807	35.5807	35.5807	35.5807	35.5807
C17	65.2796	66.1576	65.7897	66.1576	66.1576
C18	33.6706	36.5452	33.6706	36.5452	36.5452
C19	3.6283	3.8193	3.6908	3.9353	3.9353
C20	40.4052	40.4052	40.4052	40.4052	41.7519
C21	6.3425	6.3425	0	0	0
C22	9.9368	9.9368	10.4609	10.4609	15.4255
C23	24.0921	24.9986	24.4162	26.4775	26.4775
C24	22.4339	22.4953	23.7902	24.1683	27.3866
C25	19.9086	21.3887	20.0174	21.5338	25.2558
Sum	1869.351	1907.115	1414.092	1448.063	1500.848
%	24.5422	25.038	23.5741	24.1405	25.0204

Note: no grid power price for Guizhou Jinyuan (C13).

7.3.5 Conclusion and policy suggestion

In this chapter, the demonstration analysis on regulatory costs of power plants has been conducted. Regulatory costs under three situations have been discussed, which are: regulatory only for coal consumption; regulatory for coal consumption and SO_2; regulatory for coal consumption, SO_2, flue gas and NO_x. Regulatory cost of coal consumption of power plants in the PRC is about 23% of cost level in 2008; 24% for regulatory of coal consumption and SO_2; around 25% for regulatory for coal consumption, SO_2, flue gas and NO_x. In monetary, environmental regulatory cost of power plants is CNY180~200 billion (price in 2008). Meanwhile, we also calculated the inner impacts in different regulatory situations. We concluded that several points should be noticed during environmental regulatory period as below:

1) **The government should balance the environmental benefit and economic benefit during the preparation of regulatory policy.** Since 1980s, the PRC initiated the economic reforms, which brought remarkable achievements. Even though the awareness of environmental protection has been raised, conflict still exists between environmental protection and economic development. How to reasonable utilize natural resource without environmental damage is still an important issue for decision makers. The regulatory policy is the key variable for the relationship between economic development and environmental protection. Therefore, the government should balance the environmental benefit and economic benefit during the preparation of regulatory policy.

2) **Impact difference should be analyzed scientifically to coordinate interests of stakeholders before the implementation of environmental regulatory.** The regulatory policy would impact interest reallocation among different stakeholders. Reasonable cost-sharing policy could decentralize the responsibility of environmental protection. Environmental regulatory policy would impact investment direction of companies, local economic increasing, and potential environmental related parties. Preparation of regulatory policy should consider the potential impacts, to establish a rational mechanism which can coordinate interests of stakeholders.

3) **Scientific compensation mechanism should be established during the implementation of environmental regulatory to distinguish the relative impacts and absolute impacts.** In evaluating the regulatory impact of same type individual, relative impacts and absolute impacts should both be considered. Some small-scale individual body with small regulatory cost might be greatly impacted by regulatory policies. Also, for companies with large regulatory cost, how to balance the efficiency improvement and

compensation regulatory is also an important issue.

7.4 Regional scope and effect analysis of thermal power plant dispatching

7.4.1 Performance analysis method based on slacks-based efficiency measure

The performance evaluation based on slacks-based efficiency measure considers possible slack situation between input and output, thereby overcoming negative impacts due to proportional changes between input and output by DEA technology (Cook and Seiford, 2009). Slacks-based efficiency measure is widely applied in empirical evaluation of efficiency for different type of enterprises. Sueyoshi and Goto (2001) carried out the empirical research for performance of power plants in Japan between 1984 and 1993 by using the slacks-based DEA model. Liu and Tone (2008) conducted empirical analysis for banks in Japan between 1997 and 2001 by adopting three-stage DEA model. Saranga (2009) evaluated the efficiency of auto parts industry in India by using slacks-based two-stage evaluation method, and analyzed impacts of storage, technology transfer and capital management. Cheng et al.(2010) carried out empirical analysis of 34 internaitonal hotels in Taiwan and analyzed related factors which impact the performance. Sueyoshi and Goto (2011) integrated operation efficiency and environmental efficiency into DEA model, analyzed performance of fossil fuel based power plants in Japan, and found that implementation of Kyoto Protocol does not affect the efficiency of sampled enterprises during 2004~2008.

While domestic scholars have done a lot of empirical researches on environmental efficiency and economic efficiency, for example, Liu Litao and Shen Lei (2010) carried out empirical analysis on evolution of regional energy efficiency in the PRC; Qi Jianhong and Chen Xiaoliang (2011) conducted empirical analysis on relations between import/export changes and energy efficiency. However, researches on impacts analysis due to energy-efficinet power generation scheduling have not been reported yet. This paper will evaluate efficiency of 1902 thermal power plants in the PRC based on slacks-based DEA method, with consideration of different scheduling mode, energy-conservation potential and energy-conservation cost, to provide decision-making basis for further promotion and improvement of energy-efficient power generation scheduling.

7.4.2 Performance evaluation based on slacks-based efficiency measure

By using slacks-based efficiency measure proposed by Zhou et al. (2006), we calculated measurement of environmental efficiency and relative measurement of economy.

(1) Environmental economic measure

As for a typical undesirable outputs orientation DEA-based model, related environmental performance evaluation method can be:

$$PEI = \lambda^* = \min \lambda$$

$$s.t. \sum_{k=1}^{K} z_k x_{nk} \leqslant x_{n0}, n = 1, 2, ..., N$$

$$\sum_{k=1}^{K} z_k y_{mk} \geqslant y_{m0}, m = 1, 2, ..., M$$

$$\sum_{k=1}^{K} z_k u_{jk} = \lambda u_{j0}, j = 1, 2, ..., J$$

$$z_k \geqslant 0, k = 1, 2, ..., K$$

This model doesnot consider slacks situation of inputs and undesirable outputs. To overcome this shortcoming, the optimal value λ^* obtained from above equation can be introduced into following case:

$$\rho^* = \min \{ t - \frac{1}{N} \sum_{n=1}^{N} S_n^- / x_{n0} \}$$

$$s.t. \sum_{k=1}^{K} z_k x_{nk} + S_n^- = x_{n0}, n = 1, 2, ..., N$$

$$\sum_{k=1}^{K} z_k y_{mk} - S_m^+ = y_{m0}, m = 1, 2, ..., M$$

$$\sum_{k=1}^{K} z_k u_{jk} = t \lambda^* u_{j0}, j = 1, 2, ..., J$$

$$t + \frac{1}{M} \sum_{m=1}^{M} S_m^+ / y_{m0} = 1$$

$$z_k \geqslant 0, k = 1, 2, ..., K; s_n^-, s_m^+ \geqslant 0$$

Considering both environmental efficiency and economic efficiency, the EPI obtained based on slacks-based efficiency measure would be:

$$SBEI_1 = \lambda^* \times \rho^*$$

(2) Economic relative measure

Firstly, to evaluate economic efficiency of DMU_0 under undesirable outputs through following liear programming problem:

$$\theta_1^* = \min\{t - \frac{1}{N}\sum_{n=1}^{N} S_n^- / x_{n0}\}$$

$$s.t. \sum_{k=1}^{K} z_k x_{nk} + S_n^- = x_{n0}, n = 1, 2, ..., N$$

$$\sum_{k=1}^{K} z_k y_{mk} - S_m^+ = y_{m0}, m = 1, 2, ..., M$$

$$t + \frac{1}{M}\sum_{m=1}^{M} S_m^+ / y_{m0} = 1$$

$$z_k \geqslant 0, k = 1, 2, ..., K; S_n^-, S_m^+ \geqslant 0$$

When undesirable outputs are considered, economic efficiency under CRS condition can be identified as:

$$\theta_2^* = \min\{t - \frac{1}{N}\sum_{n=1}^{N} S_n^- / x_{n0}\}$$

$$s.t. \sum_{k=1}^{K} z_k x_{nk} + S_n^- = x_{n0}, n = 1, 2, ..., N$$

$$\sum_{k=1}^{K} z_k y_{mk} - S_m^+ = y_{m0}, m = 1, 2, ..., M$$

$$\sum_{k=1}^{K} z_k u_{jk} = t u_{j0}, j = 1, 2, ..., J$$

$$t + \frac{1}{M}\sum_{m=1}^{M} S_m^+ / y_{m0} = 1$$

$$z_k \geqslant 0, k = 1, 2, ..., K; S_n^-, S_m^+ \geqslant 0$$

We define another slacks-based efficiency measure for modeling environmental performance as:

$$SBEI_2 = \theta_1^* / \theta_2^*$$

Since θ_1^* and θ_2^* are respectively the economic efficiency scores when undesirable outputs are and are not considered, $SBEI_2$ could be used to model the impacts of environmental regulations on economic efficiency. Noted that $SBEI_2$ would not take any positive value larger than 1. When $SBEI_2 = 1$, θ_1^* must be the same as θ_2^*. It implies that the transformation of production process from the traditional DEA to environmental DEA technology has no effects on the economic efficiency of the DMU conserned. If $SBEI_2$ is less than one, it indicates that environmental regulations result in the waste of inputs and the loss of desirable outputs with respect to the hypothesized efficient DMU. That is to say, there is an opportunity cost due to environmental regulations. Quantitatively, the degree of regulatory impact can be measured by $1 - SBEI_2$.

7.4.3 Data source and description

All data used in this chapter is sourced from "statistic information of power industry in 2008". Due to data shortage of some power plants, we collected samples from 1902 power plants, of which, 13 from Beijing, 33 from Tianjin, 118 from Hebei, 113 from Shanxi, 264 from Shandong, 85 from Inner Mongolia, 93 from Liaoning, 44 from Jilin, 98 from Heilongjiang, 27 from Shanghai, 289 from Jiangsu, 180 from Zhejiang, 20 from Fujian, 49 from Anhui, 13 from Guangxi, 126 from Guangdong, 17 from Guizhou, 24 from Yunnan, 108 from Henan, 19 from Hubei, 43 from Hunan, 25 from Sichuan, 25 from Jiangxi, 36 from Shanxi, 18 from Gansu, 6 from Qinghai and 10 from Ningxia. We considered installation capacity as the replacement of capital, power production as desirable outputs, coal consumption as undesirable outputs. Considering the consumption of coal and related resources is inspired by Lozano and Gutiérrez (2008). Besides, due to the collected data includes power utilization rate of the plant, which can be used to calculate power consumption during power generation period. Therefore, we also considered the power consumption of plants as inputs of power production.

7.4.5 Data processing and analysis

Considered the power grid structure in the PRC, we analyzed the efficiency of power plants and energy-conservation cost in two situations. One is, energy conservation optimization through provincial power grid; the other one is, energy conservation optimization through inter-provincial power grid.

(1) Provincial scheduling based efficiency analysis

We calculated economic and environmental efficiency of independent power plant and compared their economic and environmental efficiency. We noted that $SBEI_1$ and

$SBEI_2$ evaluated the economic and environmental efficiency of power plants from two different points. Table 7-9 displays average value of efficiency of power plants based on provincial scheduling.

$SBEI_1$ includes both environmental efficiency and economic efficiency, it can be considered as a measure to evaluate economic and environmental efficiency. Meanwhile $SBEI_1$ is a standardized index. The value of average efficiency could reflect inner difference of power plants. As for the calculation results, minimum difference happened to Ningxia, Qinghai and Guizhou provinces; and maximum difference happened to Shangdong and Jiangsu provinces.

$SBEI_2$ reflects difference of regulatory cost, the average value could reflect the concentration degree of regulatory cost of regional power plants. Bigger value of efficiency index at regional level means power plants are concentrated to plants with smaller regulatory cost, which indicates that regulatory cost is even distributed. From the results of Table 7-10, if power generation is scheduled by provincial power grids, the most balanced provinces/regions in regulatory cost happened in Gansu and Tianjin. The maximum difference of regulatory cost happened in Jilin province.

Table 7-9 Average efficiency based on provincial scheduling

	λ^*	ρ^*	$SBEI_1$	θ_1^*	θ_2^*	$SBEI_2$
Beijing	0.5050	0.8423	0.4752	0.6661	0.8441	0.8044
Tianjin	0.3127	0.7827	0.2913	0.7648	0.7751	0.9870
Hebei	0.1541	0.7694	0.1464	0.6741	0.7145	0.9444
Shanxi	0.1654	0.6761	0.1408	0.5045	0.5858	0.8766
Shandong	0.0978	0.6842	0.0889	0.4692	0.5339	0.8842
Inner Mongolia	0.1641	0.6951	0.1464	0.6200	0.6556	0.9433
Liaoning	0.1591	0.70486	0.1427	0.5638	0.6456	0.8912
Jilin	0.3316	0.7700	0.3031	0.4536	0.6238	0.7638
Heilongjiang	0.2571	0.7941	0.2497	0.6457	0.7271	0.9006
Shanghai	0.4502	0.7924	0.4046	0.7057	0.8020	0.8812
Jiangsu	0.1034	0.7336	0.0940	0.5142	0.5516	0.9404
Zhejiang	0.1543	0.7740	0.1472	0.4621	0.5326	0.8930
Fujian	0.3550	0.7927	0.3316	0.6323	0.7186	0.8922
Anhui	0.3456	0.7374	0.3092	0.6732	0.7163	0.9427
Guangxi	0.5521	0.5521	0.5091	0.7623	0.8527	0.8933

	λ^*	ρ^*	$SBEI_1$	θ_1^*	θ_2^*	$SBEI_2$
Guangdong	0.2413	0.7656	0.2129	0.5892	0.6498	0.9215
Guizhou	0.7034	0.9324	0.6741	0.8580	0.8980	0.9592
Yunnan	0.4067	0.8239	0.3732	0.7504	0.8308	0.9079
Henan	0.2908	0.7319	0.2728	0.5231	0.6203	0.8700
Hubei	0.6191	0.8916	0.5812	0.8417	0.9152	0.9253
Hunan	0.3720	0.7884	0.3581	0.5635	0.7353	0.7869
Sichuan	0.4434	0.6958	0.3757	0.6660	0.7367	0.9106
Jiangxi	0.3484	0.7104	0.3298	0.6350	0.7451	0.8560
Shaanxi	0.3795	0.6855	0.3439	0.6602	0.7599	0.8719
Gansu	0.4694	0.8234	0.4197	0.8215	0.8263	0.9927
Qinghai	0.6936	0.8640	0.6827	0.8478	0.9233	0.9220
Ningxia	0.7024	0.9862	0.6982	0.8528	0.9143	0.9366

According to calculation before, we can obtain possible maximum coal-saving proportion by energy-efficient power generation scheduling through provincial power grids. Table 7-10 listed distribution of energy-conservation potential provinces and regions and estimation of energy-conservation cost by provincial scheduling. Due to the proportion is constructed based on production process of power plants, without consideration of geological difference, detailed distribution of power grids, demand-side difference and other impacts, the maximum coal-saving proportion is relatively significant. According to the calculation results, through provincial energy-efficient power generation scheduling, According to calculation results, through provincial dispatching, Inner Mongolia has the biggest energy-efficient potential, followed by Shandong, Jiangsu, Guangdong, Shanxi, Sichuan, Hebei, Henan, Tianjin, Anhui, Gansu, Jilin, Shanghai, Heilongjiang, Yunnan, Fujian, Zhejiang, Beijing, Jiangxi, Hubei, Guizhou, Guangxi, Hunan, Shanxi, Ningxia, Qinghai. Jilin province ranks number one of energy-efficient cost, followed by Zhejiang, Hunan, Henan, Guangdong, Heilongjiang, Shandong, Shanxi, Fujian, Shanghai, Ningxia, Beijing, Jiangxi, Guizhou, Liaoning, Hubei, Inner Mongolia, Jiangsu, Sichuan, Yunnan, Anhui, Hebei, Guangxi, Shanxi, Tianjin, Qinghai and Gansu. If energy-efficient optimization dispatching is conducted within provincial grid network, the ratio of energy conservation cost of thermal power plants in the PRC and national average coal consumption is 0.9816%.

Table 7-10 Distribution of energy-conservation potential provinces/regions and energy-conservation cost estimation within provincial scheduling

	Distribution proportion of energy conservation potential(%)	Coal consumption ratio of energy-efficient cost(%)
Beijing	2.8283	1.3435
Tianjin	4.1427	7.9038
Hebei	4.6844	4.6095
Shanxi	4.9358	0.8276
Shandong	5.8220	0.7975
Inner Mongolia	6.3417	2.6034
Liaoning	5.0434	1.8623
Jilin	3.6893	0.2984
Heilongjiang	3.3913	0.7246
Shanghai	3.5547	0.9207
Jiangsu	5.4579	3.0662
Zhejiang	2.9842	0.3301
Fujian	3.1287	0.8476
Anhui	4.0062	3.9635
Guangxi	2.4342	5.0532
Guangdong	5.2771	0.6983
Guizhou	2.4710	1.8593
Yunnan	3.2221	3.3482
Henan	4.6449	0.5456
Hubei	2.5902	2.5344
Hunan	2.3724	0.4913
Sichuan	4.7587	3.0778
Jiangxi	2.7879	1.5832
Shaanxi	2.4420	5.2713
Gansu	3.8701	31.7848
Qinghai	0.9362	18.9358
Ningxia	2.1829	1.1715
The PRC	100	0.9816

(2) Efficiency analysis based on regional power grid scheduling

With consideration of current grid structure in the PRC, another possible energy-efficient scheduling or generator optimization mode would be inter-regional scheduling. Table 7-11 shows distribution of regional network in the PRC. Currently, there are two major Grid Companies in the PRC: State Grid Corporation of China and China

Southern Power Grid. Generally, inter-regional power scheduling barely happens before. Table 7-12 shows average value of efficiency of

Table 7-11　Regional distribution of power grid in the PRC

Company	Regional grid	Provinces included
National Grid Company	North China Grid	Beijing, Tianjin, Hebei, Shanxi and Shandong
	East China Grid	Shanghai, Zhejiang, Jiangsu, Anhui and Fujian
	Central China Grid	Hubei, Hunan, Henan, Jiangxi, Sichuan and Chongqing
	Northeast China Grid	Liaoning, Jilin, Heilongjiang and east of Inner Mongolia
	Northwest China Grid	Shanxi, Gansu, Ningxia, Qinghai, Xinjiang and Tibet
China Southern Power Grid	China Southern Grid	Guangdong, Guangxi, Yunnan, Guizhou and Hainan

Table 7-12　Average efficiency based on regional power grid scheduling

	λ^*	ρ^*	$SBEI_1$	θ_1^*	θ_2^*	$SBEI_2$
North China Grid	0.0926	0.6943	0.0818	0.4694	0.5241	0.9033
Northeast China Grid	0.0872	0.7270	0.0787	0.3795	0.4702	0.8441
East China Grid	0.1018	0.7784	0.0952	0.4627	0.5055	0.9293
China Southern Grid	0.2281	0.7444	0.1960	0.5616	0.6153	0.9229
Central China Grid	0.2207	0.7149	0.2004	0.4791	0.5729	0.8578
Northwest China Grid	0.3560	0.8245	0.3311	0.6991	0.7517	0.9281

Through regional energy-efficient power generation scheduling, energy-efficient potential ranked as: Northeast China Grid, North China Grid, China Southern Grid, Central China Grid, East China Grid and Northwest China Grid. As for the energy-efficient cost, the ranking list would be: Northeast China Grid, Central China Grid, East China Grid, China Southern Grid, North China Grid and Northwest China Grid. If the inter-regional dispatching is carried out, the coal consumption ratio of energy-efficient cost for thermal power plants would be 0.5969%.

Through calculation, we found the ratio of maximum potential of regional energy-conservation scheduling and that of provincial scheduling is 1.3478, which means the energy-conservation potential of regional energy-efficient scheduling is bigger than that of provincial energy-efficient scheduling. As for the difference between regional energy-efficient scheduling and provincial energy-efficient scheduling, it can be concluded as below: firstly, as for energy-efficient potential, inter-regional dispatching is more possible to achieve energy conservation; secondly, as for energy-efficient cost, inter-regional dispatching increased the energy-efficient cost. The energy-efficient potential could be enhanced through regional dispatching, because power generation resources can be better optimized at a larger scaled scope. However, this optimization would increase the cost. Therefore, the decision maker has to make reasonable decision between those two modes, to achieve reasonable arrangement of resources optimization and energy-efficient scheduling planning.

Table 7-13　Distribution of energy-conservation potential and energy-conservation cost estimation through regional scheduling

	Proportion of energy conservation potential (%)	Proportion of coal consumption for energy conservation (%)
North China Grid	18.8128	0.8700
Northeast China Grid	20.7340	0.3276
East China Grid	15.7111	0.5892
China Southern Grid	17.5417	0.7516
Central China Grid	16.8199	0.5733
Northwest China Grid	10.3806	3.3732
The PRC	100	0.5969

7.4.6　Conclusion and policy recommendations

In this chapter, slacks-based efficiency measure is adopted to conduct economic environmental efficient evaluation for 1902 thermal power plants in the PRC. We have considered two power dispatching plans: one is, inter-provincial optimization dispatching of independent power plants; the other one is, inter-regional optimization dispatching of independent power plants. We calculate economic environmental efficiency of each power plant under above two situations, and gave detailed calculation results. Especially, we listed the best independent power plants under each situation. Based on above calculation, we provide ranking list of province and Grid Companies, as well as maximum potential saved energy and energy-efficient cost. Demonstration analysis was prepared under different situations.

Several points as below can be concluded during energy-efficient scheduling and optimization of thermal power plants:

(1) **To balance energy-efficient benefits and cost, to prepare optimization planning.** Through comparison the difference of energy-efficient potential and cost between provincial dispatching and regional dispatching, we found that: as for energy-efficient potential, inter-regional dispatching has greater possibility of energy-efficient; as for energy-efficient cost, the cost of inter-regional dispatching is higher than provincial one. The energy-efficient potential could be enhanced through regional dispatching, because

power generation resources can be optimized at a larger scope. However, this optimization would increase the cost. Therefore, the decision maker has to make reasonable decision between those two modes, to achieve reasonable arrangement of resources optimization and energy-efficient scheduling planning.

(2) **To distinguish energy-efficient potential and energy-efficient cost, rationalize stakeholder relationships.** No matter the provincial dispatching or the regional dispatching, energy- efficient potential and energy- efficient cost must be different from one to another. The one with great energy-saving potential is not necessarily with low cost. Therefore, during the establishment of energy conservation policy for thermal power plants, above difference must be fully considered, as well as relationships among stakeholders.

(3) **To promote energy-efficient scheduling and establish reasonable planning.** Due to great difference between dispatching agencies, with consideration of the imperfection of energy-efficient scheduling system itself, technical, environmental and economic factors of different stakeholders should be fully put into account during establishment of promotion plan for energy-efficient scheduling. Detailed plan should be established to promote energy-efficient scheduling.

8. Policy Suggestion to Improve Energy-Efficient Power Generation Scheduling

Energy conservation is the general trend of development of China's power industry and the pilot project of energy-efficient scheduling also received some success. Through the comparative analysis of the pilot program on energy generation scheduling, we found the existing problem and raise the optimization suggestion accordingly.

8.1 Strengthen management and regulatory of energy-efficient power generation scheduling

(1) **To strengthen energy statistics.** it can provide necessary basis of measurement to achieve scheduling optimization in terms of energy conservation, environmental protection and economic benefit. Besides, the amendments of market rules, establishment of energy-conservation power price and emission standards, as well as regulatory of overall energy conservation status need energy statistics and related data review. Therefore, establishment of scientific database for energy conservation is the measurement basis to implement multi-objectives in energy conservation, environmental protection and economic benefits.

(2) **To strengthen management of energy-efficient power generation scheduling.** Normally, people consider energy-efficient power generation scheduling as grid dispatching, grid connection, construction of information system and other technical issues. However, those are only facial issues. There are three issues for system construction of energy-efficient power generation scheduling. The first issue is grid connection of computers, construction of real-time online system for coal consumption and flue gas desulfurization. This is the primary objection and easily achieved through financial investment. Experts can ensure the achievement of this issue. Also, it contains the big proportion of financial investment. The second issue is organization during development procedure, not only including personnel, but also allocation of users. It is a management issue of energy-efficient power generation scheduling project. It needs to achieve the balance of personnel, financial and material issues. The third level is to construct reasonable structure and new operation mechanism of energy-efficient power generation

scheduling. This is important to achieve replacement of the old scheduling system, and realize the fundamental objective of energy-efficient power generation scheduling and actual benefits from new scheduling rule. Therefore, the implementation of energy-efficient power generation scheduling for power grid in the PRC is basically a management reform. It is easy to invest and construct system equipments. However, to achieve energy conservation, management is essential.

(3) **To devote increasing efforts for information disclosure and strengthen the regulatory of the energy-efficient power generation scheduling.** Related agencies should provide information needed for energy-efficient scheduling timely, accurately and completely to the appropriate as required. The supervision, statistics, analysis of relevant information and regular report to the public should be conducted on the execution of energy-saving program, to maintain the legitimate rights and interests of market players.

8.2 Establish market mechanism for energy-efficient power generation scheduling

We should continuously establish regional power market system, to provide favorable power supply environment for EEPGS implementation. Construction of regional power market is helpful for optimization of industrial structure and guiding reasonable investment to allocation power resource at a larger scale.

(1) **To revise power market rules to meet requirements of energy-efficient power generation scheduling.**

When the power market pricing mechanism is not well developed and the energy-efficient environmental-protection power price is not established, auction price should be revised according to not only financial cost, but also energy consumption per unit, pollutants emission index per unit, transportation cost, network line loss and other factors. In this way, we pay fully attention on economic, energy-saving and environmental protection issues to achieve energy conservation and emission reduction, and transfer integrated benefits to the whole society. On-grid power price formation mechanism adapted for EEPGS should be gradually established to further promote power market construction.

(2) **To research on provincial and regional transmitted power under EEPGS mode.**

Under EEPGS mode, large units ranking in the top of the sorting table have priority of power generation. Thereby they would be fully used, in this way, they would have less competition in surplus power transmission motivation. In areas with large proportion of units of 300 MW, some units might have certain idle power amount. The surplus power volume can be transmitted to other provinces or regions. However, with limitation principle of "utilization density of units with capacity of 300 MW should be less than units with capacity of 600 MW", the surplus power volume can be transmitted is limited. Units with capacity of 200MW is not allowed to be transmitted under EEPGS rules. Impacted by above factors, after full implementation of EEPGS, it will be difficult to organize surplus power transmission to other areas. With decrease of transmitted power volume, we suggest to fully implement the LSS program to give more power generation quotas to large units through auction, market-matching, bilateral negotiated transaction and other methods. In this way, it is not only resolved the compensation issue for small units, but also helpful to achieve energy conservation and emissions reduction.

(3) **To establish large-scale comprehensive power transaction center**

National, regional and provincial power transaction agency should be established for public listing transaction of active power, power volume for ancillary services, renewable energy and carbon index, to reduce regional difference and balance parity system of power resource. Shadow price and opportunity cost of investment in power source, power grid and electric fields can only be calculated through market mechanism. The adjustment signal of energy conservation can be conveyed through pricing system to improve speed, direction and scale of resource optimization allocation. The power regulatory can only be smoothly conducted through independent transaction function. Only in this way, the power regulatory could step into a sound development way. Also, independent transaction can explain the origins of the loss of power plants, to distinguish process cost responsibility and identify the influential factors on loss, to eventually promote healthy and stable development of power industry and give clear answer to the society on power cost, benefits of power enterprises and state-owned management performance.

8.3 Promote development of energy-efficient power generation by using economic approach

(1) **Improve the economic compensation approach.** Energy-efficient power generation scheduling is a complicated systematic project covering different aspects. The promotion difficult is to balance interests of multi-level stakeholders. In order to encourage contribution of high-efficient units to energy conservation, appropriately extension of benefits of high-efficient units is necessary. But large amount of small units have been idle and face a serious issue that they might be closed. Enhanced power purchase fee for power grid companies should be compensation appropriately too. As for power plants, reasonable

compensation period should be established for small thermal units to be decommissioned. During the compensation period, certain power generation quotas should be allocated for them. Compensated power generation quotas for decommissioned small thermal units should be transferred to large units with agreed price. This part of income can be used for production switches and employee resettlement. For pumped storage power station providing peak shaving and frequency regulation functions, they can be rented. The calculation of compensation fee for backup units should be simplized; power producer should be responsible for equipments maintenance to ensure reliable operation; power grid is responsible for scheduling utilization of units. According to the available capacity, power grid company pay certain rental fee for power producer according to the signed contract. The rental fee is not related to actual power generation volume. Most self-supply power plants should be replaced by big power grid enterprises due to its small capacity, backwards techniques, high coal consumption, high pollution without desulfurization facilities. As for power grid, the implementation of EEPGS enhanced the power purchase cost and power transmission/distribution grid transformation. The incremental fee can be compensated through special national funds used for energy conservation and environmental protection.

(2) **Promote energy conservation and emission reduction in power industry through fiscal and tax policy.** The government issued related compensation policy to cover the cost by EEPGS; related tax policy which supports energy conservation and emission reduction should be further improved; tax preferential policy should be carried out for utilization and development of renewable energy including biomass, geothermal, solar resources; for high-emission and high-pollution enterprises, environmental protection regulatory should be strengthened, and related waste discharge tax should be collected. Pollutants emission index and water use permission index transaction should be conducted. Pollutants emission and water use permission index of decommissioned units can be traded to gain certain economic compensation.

(3) **Establish energy-efficient incentives in accordance with market economy requirements.** We should fully take advantage of the function of market in efficiency improvement. We should establish related pricing, financial, tax and credit policies which have positive impacts on energy saving. This is the necessity to guide and promote the social energy saving activities. Especially, a rational energy price policy is significant important. Price is the basic method to allocate resource through market mechanism. If the energy price is irrational, various resource could not be allocated in a energy-conservation way. We should pay more attention on adjustment of energy price system and reform of energy product price formation mechanism. We should further study on the feasibility of

energy consumption tax. We should prepare energy-efficient product list to reduce related tax for products in this list. Related awarding measures should be carried out for enterprises adopting advanced, high-efficient and energy-conservation equipments. We should direct related banks to provide subsidized loans. We should also direct commercial banks to provide energy-conservation loans, and encourage local government to establish energy-conservation development special funds to support energy conservation technology development and establishment of energy-efficient policies. We should continue to support independent research and development of advanced energy-efficient technologies and products applicable to enhance the overall level of Chinese industrial energy-saving technology.

8.4 Policy suggestions on actively promoting power reform in a stable manner

At present, under the background of climate change, the problems in terms of resources, energy and environment are getting more serious along with the development of economy in the PRC. Development of a low-carbon economic development path is a necessary requirement of scientific development to promote energy. Low-carbon development is also a new requirement proposed for power system reform. Therefore, we should fully carry out the scientific development theory, change the increasing way of power industry, take advantage of good timing for resolve power demand-supply conflict, learn successful reform experience from foreign countries to solidate the reform achievements and stably promote the construction of power market.

(1) **Smoothen power price formation mechanism and speed up power price reform.** Under current power regime, power transmission and distribution price should be determined according to different voltage grade at provincial level based on the current difference between power purchasing price and power sale price. The power transmission price of inter-regional and inter-provincial transmission should be determined at regional level. The power price for major users should be gradually established based on related policy on rural and residential power price.

(2) **Take advantage of EEPGS to speed up construction of power market platform.** In areas with certain condition, two-settlement price can be applied to implement energy-efficient power generation scheduling. Meanwhile, capacity market, bilateral transaction market and market for ancillary service should be established to provide integrated, fair and open power market.

(3) **Promote power transmission and distribution system, dispatching transaction**

agency and rural power regime reform. We should actively explore practical form to separate power transmission and power distribution, ensure the independent of power dispatching transaction, change current financial investment system, carry out franchise operation of power plants, promote diversity of ownership of power distribution, encourage diversity of power supply and power selling companies and coordinate even development of rural and urban power grid.

(4) **Speed up reform of ownership system, further open access permission.** We should encourage investment bodies, including private and foreign capital to invest power plants and power distribution companies. Overall reform to power generation enterprises should be carried out to reconstruct structure. Reform for "tertiary and multi-operation" enterprises should be carried out through fair market competition to promote healthy development of power enterprises.

(5) **Strengthen reform tensity of governmental agencies and speed up power legal system construction.** We should study on and establish energy management department or energy regulatory agency integrating administrative and regulatory function, to fulfill responsibilities of government and regulatory agency and reform even-load power generation quotas allocation. In this way, the market can play the role of resource allocation function. And government can play power planning and technology management functions as well. Power legal system construction should be speeded up to improve healthy power market regulatory system and implement effective regulatory based on related laws and regulations.

References

[1] А. С. Горшков, 1991. Technical economy of thermal power plants. (translated by Lin Qihua and etc.). Beijing: China Water Power Press.

[2] DL/T 783-2001. 2001. Guideline of water-saving for thermal power plants. Beijing, China Electric Power Press.

[3] Fiona Woolf, Global power transmission expansion—the path to success, China Electric Power Press, Beijing.

[4] GB/T 15587-1995, 1995. Guideline of energy management of industrial companies. Beijing: China Standard Press.

[5] GB17167-2006 17167, 2006. Requirement for metrology management of energy consumption. Beijing: State General Administration of Quality Supervision, Inspection and Quarantine.

[6] Bai Xuejie, Song Ying, 2009. Environmental specifications, technical innovation and efficiency improvement of thermal power industry in China, China Industry Economy, 8, 68~77.

[7] China Electric Power Encyclopedia. 1995. Beijing. China Electric Power Press.

[8] Chen Yuan. 2007. Energy safety and energy development strategy research. Beijing. China financial economic press.

[9] Chi Yuanying, Wang Yanliang, Niu Dongxiao and etc, 2010. Optimization mode of generation right trading under carbon emission trading. Grid technology, 6, 78~81.

[10] Dong Jun, Ma Bo, Zhou Xiang. 2008. Discussion on impacts of energy-efficient power generation scheduling to power plants and compensation mechanism. Power technology and economy, 20(6): 52~56.

[11] Implementation program of energy-efficient power generation scheduling pilot project in Guangdong Province, 2007.

[12] Implementation program of energy-efficient power generation scheduling pilot project in Guizhou Province, 2007.

[13] State electricity regulatory commission. Notice regarding on pilot program submission of regulation of direct trading between power plants and power users.www.serc.gov.cn/wdd/20100317_12779.htm

[14] State electricity regulatory commission, 2006. Power reform overview and power regulatory capacity building, Beijing, China Water Conservancy and Hydropower Press.

[15] State electricity regulatory commission, 2006. power industry market reform in South America, Asian and African countries, Beijing, China Water Conservancy and Hydropower Press.

[16] State electricity regulatory commission, 2006. Power market in European countries and Australia, Beijing, China Water Conservancy and Hydropower Press.

[17] State electricity regulatory commission, situation announcement of energy conservation of power plants in 2009, 2010.

[18] State electricity regulatory commission, situation announcement of energy conservation of power plants in 2008, 2009.

[19] State electricity regulatory commission, 2009. Explore of direct power purchase by major users. Beijing, China Electric Power Press.

[20] State Grid, 2010. SGCC Green Development White Paper.

[21] The National Development and Reform Commission, 2007. Integrated working program of energy cosnervation.

[22] The National Development and Reform Commission, 2007. The 11th Five-year plan of energy development.

[23] The National Development and Reform Commission, 2004. Medium/long term specific planning on energy conservation.

[24] The National Development and Reform Commission, 2005. Energy conservation experience of developed countries. China Economic & Trade Herald, 14, 13~17.

[25] National Bureau of Statistics of China. China Energy Statistic Yearbook. China Statistic Publication Press.

[26] State Power Economic Research Institute, 2009. Annual analysis report for foreign power market reform(2008). Beijing: China Electric Power Press.

[27] Implementation program of energy-efficient power generation scheduling pilot project in Henan Province, 2007.

[28] Huang Yonghao, Shang Jincheng. 1999. Research on the operation mode of power market and technical supporting system. Beijing, Science Press.

[29] Implementation program of energy-efficient power generation scheduling pilot project in Jiangsu Province (draft), 2007.

[30] Jeremy. D. Lambert, Hu Jiangyan translated, 2007. PJM power market in USA, China Water Conservancy and Hydropower Press.

[31] Jin Pei, 2009. Theory research on relationship of resource environmental regulatory and industrial competition, China Industrial Economy, 3, 5~17.

[32] Jin Sanlin, 2007. Trends and countermeasures of energy conservation in China, Economic Review, 5, 35~35.

[33] Li Chanbin, Kang Chongqing, Xiaqing and etc, 2003. Power generation right trading and the mechanism analysis, Automation of Electric Power System, 6, 13~18.

[34] Li Jinchao, 2009. Research power optimal operation based on energy-efficient scheduling and demand side management, doctoral thesis of North China Electric Power University.

[35] Li Shuhui, Li Guanghuai, Li Shiying, etc,. 2008. Existing problems and suggestions for power grid operation in energy-efficient power generation scheduling, Jilin Power, 36(6): 54~56. 2008.

[36] Liu Litao, Shen Lei, 2010. Temporal and spatial evolution and its impact factors analysis for China's regional energy efficiency, Journal of Natural Resources, 25(12), 2142~2153.

[37] Progress report of China Southern Power Grid's power industry development 12th Five-year plan and middle-long term planning, 2009.

[38] Documents from Communication meeting of China Southern Power Grid energy-efficient power generation scheduling pilot project, 2009.

[39] Ni Xuelin, 2005. Technical economic index calculation of thermal power plants. Beijing. China Water Power Press.

[40] Peng Mengyue, 2007. Financing mechanism for energy management in US-ESPC. Construction technology, 24, 6~7.

[41] Qi Jianhong, Chen Xiaoliang, 2011. Import/expoert and energy utilization efficiency: demonstration research based on panel data of Chinese industry, Southern Economy, 1, 14~25.

[42] Qian Bozhang, 2008. Energy conservation—necessary path to sustainable development. Beijing: Science Press.

[43] Sally Hunt, Yi Liyun translated, 2004. Power market competition, Beijing: Zhongxin Press.

[44] Shang Jincheng, Zhang Liqing, 2007. Research and application of energy conservation in power industry and resource optimization technology, Grid Technology, 22, 58~63.

[45] Shang Jincheng, 2009. Theory and application of power generation rights based energy conservation (2), power system automation, 13, 37~42.

[46] Shang Jincheng. 2009. Research on economic compensation mechanism of energy-efficient power generation scheduling (2): market mechanism based economic compensation design and analysis, power system automation, 33(3)：46~50.

[47] Shang Jincheng. 2009. Research on economic compensation mechanism of energy-efficient power generation scheduling (1): market mechanism based economic compensation design and analysis, power system automation, 33(2): 44~48.

[48] Shang Jincheng, 2009. Theory and application of power generation rights based energy conservation (1), power system automation, 12. 46~52.

[49] Si Dan and etc, 2011. Function evaluation and model research of power grid in construction of low carbon energy system, research report of Industrial economy devision of CASS.

[50] Implementation program of energy-efficient power generation scheduling pilot project in Sichuan Province, 2007.

[51] Wang Bing, Wu Yanrui, Yan Pengfei, 2008. Environmental regulatory and total factor productivity grouth: empirical research of APEC, Economic research, 5, 19~32.

[52] Wu Yulin, Wen Fushuan, Ding Jianying and etc, 2010. Bidding strategy discussion of power plants in power generation right trading market, Automation of Power System, 17, 6~12.

[53] Xu Rong, Zhao Yan, Li Lei and etc, 2007. Benefits analysis of energy conservation based energy generation right trading, Hydropower energy science, 6, 150~153.

[54] Xu Shichun, He Zhengxia. 2007. Environmental regulatory, product quality and corporate revenue—utilization efficient of pollution fines policy, Finance and Trade Economics, 3, 1~9.

[55] Yang Hongliang, Shi Dan, Energy efficiency research methods and comparison of regional energy efficiency in China, Economic theory and economic management, 2008, 3, 12~20.

[56] Yang Hongliang, Shi Dan, Xiao Jie, Impacts of natural factors to energy efficiency—potential and actual energy saving in China, Industrial economy in China, 2009, 4, 73~84.

[57] Yang Kun, Sun Yaowei, Liang Zhihong, 2007. Power market and targeting mode. Beijing, China Electric Power Press.

[58] Training workshop of former State Economic and Trade Commission to US, 2003. Enlightenment of energy conservation policy and management mode in US, Energy conservation and environmental protection, 8, 1~5.

[59] Zhang Baoguo. 2010. China Energy Development Report 2009. Beijing: Economic Science Press.

[60] Zhang Senin, Chen Haoyong, Qu Shaoqing and etc, 2010. Bilateral trading in the power market and benefit analysis of energy conservation, East China Power, 3, 0332~0336.

[61] Zhang Shishuai, Zhang Xuesong, Wang Wen and etc, 2010. Energy conservation indicator design and application analysis under power generation trading, Grid Technology, 11, 156~161.

[62] Zhao Zunqian, Xin Yaozhong, Guo Guochuan and etc, 2001. Power market operation system. Beijing: China Power Press.

[63] China Southern Power Grid, 2007. Annual summary of scheduling operation in 2007, 2009.

[64] The State Council Information Office, 2007. China's energy situation and policy.

[65] Zhu Chengzhang, 2007. Objective understanding of the grim situation of China's energy conservation, Sino-Global Energy, 12, 13~19.

[66] Yu Erjian, 1986. Economic scheduling of modern power system. Beijing: China Water and Power Press.

[67] Yanyu, Make, Yuzhao, 2007. Discussion on improvement of power generation scheduling

mode to achieve energy conservation, environmental protection and economic dispatching. China Power, 40(6), 6~9.

[68] Zhang Lizi, Xie Guohui, Huang Renhui, 2008. Concept of post power market reform in China. Journal of North China Electric Power University, 35(6), 17~20.

[69] Xie Guohui, Zhang Lizi, Shu Jun, 2009. Energy-efficient power generation scheduling mode under market mechanism, Automation of Power System, 33(8), 29~32.

[70] Zhang Zhigang, Xiaqing, 2009. Framework and key technologies of power generation scheduling of intelligent power grid, Grid Technology, 33(20), 1~8.

[71] Zhuangyan, Jiang Liping, Ma Li and etc, 2010. The content, comparison and relationship between power generation right trading and emission right trading, China Power, 1, 5~8.

[72] Zhou Jiquan, 2008. Clean production to promote energy conservation, energy conservation and environmental protection, 24, 28~28.

[73] Abbott, M., 2006, The productivity and efficiency of the Australian electricity supply industry. Energy Economics 28, 444~454.

[74] Boyd, G., Tolley, G., Pang, J., 2002. Plant level productivity, efficiency, and environmental performance of the container glass industry, Environmental and Resource Economics, 23, 29~43.

[75] Brännlund,R., Färe, R., Grosskopf, S., 1995. Environmental regulation and profitability: an application to Swedish pulp and paper mills, Environmental and Resource Economics, 6, 23~36.

[76] Chang, Y. C., Wang, N., 2010. Environmental regulations and emissions trading in China, Energy Policy, 38, 3356~3364.

[77] Cheng, H., Lu, Y., Chung, J. 2010. Improved slacks-based context-dependent DEA-A study of international tourist hotels in Taiwan, Expert systems with Applications, 37, 6452~6458.

[78] Cook, W. D., Seiford, L. M., 2009. Data envelopment analysis (DEA)-Thirty years on, European Journal of Operational Research, 192, 1~17.

[79] Deng M. 2009. 6th International Multi-Conference on Systems, Signals and Devices, 212~218.

[80] Färe, R., Grosskopf, S., Lovell, C. A. K., Pasurka, C., 1989. Multilateral productivity comparisons when some outputs are undesirable: a nonparametric approach. The Review of Economics and Statistics 71, 90~98

[81] Färe, R., Grosskopf, S., Lovell, C. A. K., Pasurka, C., 1989. Multilateral productivity comparisons when some outputs are undesirable: a nonparametric approach. The Review of Economics and Statistics 71, 90~98

[82] Farrell, M., 1957, The measurement of productive efficiency. Journal of the Royal Statistics Society. Series A 120, 253~282.

[83] Feng Zhijun, Yan Nailing. 2007. Putting a circular economy into practice in China. Sustain, 2, 46~56.

[84] Gao, C., Li, Y., 2010. Evolution of China's power dispatch principle and the new energy saving power dispatch policy, Energy Policy, 38, 7346~7357.

[85] GAO. 2005, Meeting Energy Demand in the 21st Century (Z). GAO Report.

[86] Golany, B., Roll, Y., Rybak, D., 1994. Measuring efficiency of power plants in Israel by data envelopment analysis. IEEE Transactions on Engineering Management 41, 291~301.

[87] Grob, Gustav R. 2009, Future transportation with smart grids and sustainable energy.

[88] IEA. 2007. World Energy Outlook 2007, OECD/IEA. Paris.

[89] Jamasb, T., Nillesen, P., Pollitt, M., 2004. Strategic behavior under regulatory benchmarking. Energy Economics 26, 825~843.

[90] Jamasb, T., Pollitt, M., 2003. International benchmarking and regulation: an application to European electricity distribution utilities. Energy Policy 31, 1609~1622.

[91] Lam, P. L., Shui, A.,2001, A data envelopment analysis of the efficiency of China's thermal power generation, Utilities Policy, 10, 75~83.

[92] Liu, J., Tone, K., 2008, A multistage method to measure efficiency and its application to Japanese banking industry, Social-Economic planning Sciences, 42, 75~91.

[93] Lozano, S., Gutiérrez, E., 2008. Non-parametric frontier approach to modeling the relationships among population, GDP, energy consumption and CO_2 emissions. Ecological Economics 66, 687~699.

[94] Lozano, S., Gutiérrez, E., 2008. Non-parametric frontier approach to modeling the relationships among population, GDP, energy consumption and CO_2 emissions. Ecological Economics 66, 687~699.

[95] Lozano, S., Gutiérrez, E., 2008. Non-parametric frontier approach to modeling the relationships among population, GDP, energy consumption and CO_2 emissions. Ecological Economics 66, 687~699.

[96] Min Y. 1997. The German Circular Economy Act, Environ Guiding News, 3, 11~18.

[97] Nag, B., 2006. Estimation of carbon baselines for power generation in India: the supply side approach. Energy Policy 34, 1399~1410

[98] Olatubi, W. O., Dismukes, D. E., 2000. A data envelopment analysis of the levels and determinants of coal-fired electric power generation performance. Utilities Policy, 9, 47~59.

[99] Pasurka, C. A. 2001, Technical change and measuring pollution abatement cost: an activity analysis framework, Environmental and Resource Economics, 18, 61~85.

[100] Pombo, C., Taborda, R., 2006. Performance and efficiency in Colombia's power distribution system: effects of the 1994 reform. Energy Economics 28, 339~369.

[101] Ren Yong. 2007. The circular economy in China. Master Cycles Waste Manage, 9, 45~55.

[102] Saranga, H., 2009. The Indian auto component industry-Estimation of operational efficiency and its determinants using DEA, European Journal of Operational Research, 196, 707~718.

[103] Sueyoshi, T., Goto. M., 2001. Slack-adjusted DEA for time series analysis: Performance measurement of Japanese electric power generation industry in 1984~1993, European Journal of Operational Research, 133, 232~259.

[104] Sueyoshi, T., Goto. M., 2011, DEA approach for unified efficiency measurement: Assessment of Japanese fossil fuel power generation, Energy Economics, 33, 292~303.

[105] Thakur, T., Deshmukh, S. G., Kaushik, S. C., 2006. Efficiency evaluation of the state owned electric utilities in India. Energy Policy 34, 2788~2804.

[106] UK Government. 2003. Energy White Paper, Our Energy Future: Creating a Low Carbon Economy.

[107] UNIDO. 2008. Powering Industrial Growth-The Challenge of Energy Security for Africa. Working Paper.

[108] UNIDO. 2008. Powering Industrial Growth-The Challenge of Energy Security for Africa working Paper.

[109] Vardanyan, M., Noh, D., 2006. Approximating pollution abatement costs via alternative specifications of multi-output production technology: a case of the US electric utility industry. Journal of Environmental Management, 80, 177~190.

[110] Xia, X. H., Huang, G. T., Chen, G. Q, Zhang, B., Chen, Z. M. Yang, Q., 2011. Energy security, energy efficiency and carbon emission of Chinese industry, Energy Policy, Forthcoming.

[111] Yaisawarng, S., Klein, J. D., 1994. The effects of sulfur dioxide controls on productivity

change in the U.S. electric power industry. The Review of Economics and Statistics 76, 447~460.

[112] Yang Hongliang, Pollitt Michael, 2009, Incorporating both undesirable outputs and uncontrollable variables into DEA: The performance of Chinese coal-fired power plants, European Journal of Operational Research, 197, 1095~1105.

[113] Yang Hongliang, Pollitt Michael, 2010, The necessity of distinguishing weak and strong disposability among undesirable outputs in DEA: Environmental performance of Chinese coal-fired power plants, Energy Policy, 38, 4440~4444.

[114] Zhou, P., Ang, B. W., Poh, K. L., 2008. Measuring environmental performance under different environmental DEA technologies. Energy Economics 30, 1~14.

[115] Zhou, P., Ang, B. W., Poh, K. L., 2006. Slacks-based efficiency measures for modeling environmental performance. Ecological Economics, 60, 111~118.

[116] Zhu D. 1998. The circular economy and Shanghai's countermeasures. Social Sci, 10, 35~45.

This book is published with financial support from Innovation Project of CASS.

This book is the result of a co-publication agreement between Social Sciences Academic Press (China) and Paths International Ltd.

Power Generation in China: Research, Policy and Management
Chief Editors: Shi Dan, Yang Hongliang

ISBN: 978-1-84464-335-6

Paths International Ltd.

Printed in the United Kingdom

www.pathsinternational.com

CPSIA information can be obtained at www.ICGtesting.com
Printed in the USA
BVOW08*0004070114

341078BV00006B/26/P

9 781844 643356